Lena Kästner
Philosophy of Cognitive Neuroscience

Epistemic Studies

—

Philosophy of Science, Cognition and Mind

Edited by
Michael Esfeld, Stephan Hartmann, Albert Newen

Volume 37

Lena Kästner

Philosophy of Cognitive Neuroscience

Causal Explanations, Mechanisms, and Experimental Manipulations

DE GRUYTER

ISBN 978-3-11-065391-5
e-ISBN (PDF) 978-3-11-053094-0
e-ISBN (EPUB) 978-3-11-052920-3
ISSN 2512-5168

Library of Congress Cataloging-in-Publication Data
A CIP catalog record for this book has been applied for at the Library of Congress.

Bibliographic information published by the Deutsche Nationalbibliothek
The Deutsche Nationalbibliothek lists this publication in the Deutsche Nationalbibliografie;
detailed bibliographic data are available on the Internet at http://dnb.dnb.de.

www.degruyter.com

to my parents
who always supported me
although they never really knew what I was doing

Preface

This book is about how scientists, especially cognitive neuroscientists explain phenomena. Almost six years have passed between drafting its first pages and the final publication. In March 2014, I submitted an earlier version of the manuscript as a PhD dissertation to the Fakultät für Philosophie und Erziehungswissenschaft at Ruhr-University Bochum, Germany. I defended six it months later and was awarded *summa cum laude*. For various reasons, the publication process took much longer than anticipated. The core themes of this book, however, continue to take center stage in contemporary philosophy of science.

This is especially true for discussions surrounding *mechanistic constitution* relations and the use of *interventions* to uncover them. Mechanistic constitution is a special kind of part-whole relation that obtains between a phenomenon to be explained (the behavior of the mechanism as a whole) and the acting entities in the mechanism implementing the phenomenon in question (see chapter 3.1). Interventions are manipulations of some factor X with respect to some other factor Y that, provided appropriate conditions hold, can be used to detect a *causal* relation between X and Y. The basic idea is that if we can "wiggle" Y through "wiggling" X, we can infer that X causes Y. However, for this reasoning to work it must be possible, at least *in principle*, to independently manipulate X and Y, viz. "wiggle" X without also "wiggling" Y. But if X and Y stand in a part-whole relation this is not the case. Therefore, using interventions to detect mechanistic constitution relations is highly problematic.

Many papers now central in the debate about interventions in mechanisms have already been circulating in the community while I was working on the dissertation in Bochum. For this publication, the references have been updated accordingly and more recent work on the topic has been included in the discussion. This applies especially to chapters 6, 7, and 8. Meanwhile, a number of authors have confirmed the systematic problems I identified for using interventions in mechanistic contexts. And, moreover, that the same systematic problems arise when trying to apply interventions to the problem of *mental causation*. However, despite vigorous discussions in the philosophy of science ommunity, no unequivocal solution to these problems has yet been found. In chapter 11, I suggest a way of handling interventions in scenarios where non-causal dependence relations (such as mechanistic constitution or supervenience) are present. The flipside of my proposal is that interventions cannot stand alone to discover different kinds of dependency relations; other experiments—using *mere interactions* (see chapter 12) rather than interventions—are indispensable.

DOI 10.1515/9783110530940-203

Therefore, a second core topic discussed in this book is which role *different kinds of experiments* play throughout the *discovery* process as well as the subsequent construction of scientific *explanations* based on the discoveries made. I call the branch of philosophy of science concerned with questions of experimentation and the epistemology of scientific experiments *philosophy of experimentation*. Though the nature of scientific experiments has been discussed by philosophers at least since Francis Bacon's *Novum Organum* (1620), experimentation has not been a dominant topic in philosophical discussions on explanation. Rather, *logical positivists* have emphasized the formal characteristics of explanation, *laws of nature*, and *deduction*. More recently, such views of scientific explanation have been replaced by *causal-mechanical* accounts. They emphasize that scientific explanations aim to uncover the *causal nexus* of the world to render observed phenomena (the explananda) intelligible. This is the context in which contemporary mechanistic and interventionist accounts arise. But still, very little is said about where the relevant evidence for scientific explanations comes from.

According to our common contemporary understanding, experimentation is at the heart of scientific practice. Experiments clearly are an indispensable part of scientists' activities across a vast range of disciplines. Thus, experiments vary significantly depending on the research area, available tools and methodologies, as well as the concrete research question at hand. Despite (or precisely because of) this variety, scientific experiments enable discoveries, help generate evidence, and eventually fuel scientific explanations. But what exactly *are* experiments and what do they *actually* tell us? If we want to truly understand scientific explanations, these are the questions that need answering. As a first step into this direction, I analyze various types of experiments with respect to their *designs*, what kinds of *manipulations* they employ, and what *research questions* they aim to answer. As a result, I provide a *catalog of experiments* in chapter 14.

The point of my catalog of experiments is not merely to sort experiments into classes. Much more importantly, it is to call attention to the empirical reality of experimental research practice. Philosophers aiming to provide an empirically adequate account of scientific explanation ought to take into account a whole range of different experimental strategies that are applied at different stages throughout the discovery process. In this context, especially the introduction of *mere interactions* and *pseudo-interventions* provides an innovative and pioneering contribution to contemporary philosophy of science that helps reconfigure our understanding of how experiments reveal the world. As we think beyond (but do not neglect) intervention-based experimentation we achieve a fuller and more realistic conception of how scientists explain phenomena. While my focus in this book are explanations in cognitive neuroscience, much of what we learn from this paradigm case of an *interlevel special science* will generalize to other special sciences where we find explanations spanning multiple levels or scientific domains.

This project tremendously benefitted from the help and support I received during its various stages. I am grateful to everyone who contributed their thoughts over coffee, commented on my work at conferences, provided detailed feedback on earlier versions of individual chapters, joined me in extensive discussions of different themes covered in this book, or provided personal support.

While it is impossible to provide a complete list of names here, I would like to express my special thanks to Albert Newen and Carl Craver for supervising me during my PhD. I am grateful to Bill Bechtel and Carl Craver for inviting me to visit them at UCSD and WUSTL in summer 2013, respectively. It was during this summer that my thoughts on mere interactions took shape, especially during bike rides through Forest Park with Carl. Exchanges with Marie Kaiser, Beate Krickel and Dan Brooks in the Levels-Reduction-Mechanisms (LeRM) group have also been particularly encouraging.

While writing the major parts of this book Anna Welpinghus, Tanya Hall, To-moo Ueda, Andrea Kruse, and Sanneke De Haan have been truly great office mates who listened, discussed, and ensured I did not grow too lonely. For ample discussion and helpful comments on my work I further thank Leon de Bruin, Peter Brössel, Raphael van Riel, Tom Polger, Vera Hoffmann-Kolss, Michael Baumgartner, John Bickle, Markus Eronen, Jens Harbecke, Alex Marcellesi, Dan Burnston, Ben Sheredos, Casey McCoy, Felipe Romero, Mike Dacey, Alex Reutlinger, Rebekka Hufendiek, Victor Gijsbers, Teresa Behl, Pascale Ruder, Francesco Marchi, Lise Marie Andersen, Carrie Figdor, Philipp Haueis, Astrid Schomäcker, Lara Pourab-dolrahim, Irina Mikhalevich, David Colaco, Henrik Walter, Rob Rupert and Bob Richardson.

Outside academia, I am very happy to have family and friends who do not care about logic, causation, or experimental designs but simply get me back down to earth—well, or up a wall. My very special thanks to Monika Segelbacher for count-less spirit-raising pigeon stories, as well as to Mette Rathjen, Frank Wolter, Tanja Füth, and Christiane Mietzsch. And to my parents, who supported me unquestion-ingly in doing something that never really made sense to them: *danke euch!*

For helping with the final touches of the book, I am further grateful to Damian Warnke, Julia Herrmann, Meike Wagner, Tobias Mayer, Michael Brändel and Nadja El Kassar.

For their financial support I am indebted to the Studienstiftung des Deutschen Volkes e.V. and Ruhr-University Research School as well as the Berlin School of Mind and Brain at Humboldt-Universität zu Berlin.

Contents

Part I: Stage Setting

List of Figures

DOI 10.1515/9783110530940-205

1 Introduction

1.1 Agenda of This Book

This book is about *how scientists explain cognitive phenomena*.[1] The account I develop is tailored to scientific explanation in cognitive science, especially cognitive neuroscience. My project is primarily *descriptive* rather than normative in nature. I set out from the success stories of empirical research into memory and language processing, respectively. By systematic experimentation, scientists have made substantial progress on explaining these phenomena. My aim is to examine the research strategies behind this success and reflect them against the background of contemporary philosophy of science. I will draw two major conclusions. First, although often treated as a 'package deal' the most promising accounts of scientific explanation to date (mechanistic explanations and interventionism) are incompatible. Second, even if we succeed in rendering them compatible, they take into account only a very limited portion of the research practices scientists actually use to investigate cognitive phenomena. To understand the epistemology of scientific experimentation we must consider a wider range of experimental designs, however. This includes various observational and manipulative techniques scientists employ as well as different ways of data analysis. Taking a first step into this direction, I develop a *catalog of experiments*. This catalog characterizes different studies with respect to what kinds of manipulations they employ and which sets of research questions they can potentially answer. Along the way I will discuss how to best conceive of interlevel relations and how to modify interventionism such that it can be used to uncover different dependence relations.

1.2 Synopsis

Already in the 17th century, Francis Bacon realized that "the secrets of nature reveal themselves more readily under the vexations of art than when they go their own way." (Bacon 1960 [1620], XCX) Merely observing an object or phenomenon, that is, is not as telling as inspection under manipulation. This principle is the very backbone of empirical research: when trying to understand and explain how a phenomenon comes about, scientists systematically manipulate it. But what

---

1 This is *not* a book about metaphysics. When talking about "processes", "capacities", "phenomena", "events", and the like I use these notions in a non-technical sense unless otherwise indicated.

DOI 10.1515/9783110530940-001

exactly is the logic behind these manipulations? How do we get from manipulations to explanations? What inferences can we make based on which experiments? These are questions about the epistemology of scientific experiments. They fall within what I call *philosophy of experimentation*.

Philosophy of experimentation is a division of philosophy of science. Contemporary philosophy of science has focused on one particular kind of manipulation in the context of explanations: manipulations of some factor Y through interventions I on X. While such *interventions* have been discussed extensively (e.g. Spirtes et al. 1993; Pearl 2000; Woodward 2003, 2008a; Reutlinger 2012), different authors have presented slightly different conceptions. The most popular account is arguably James Woodward's (2003; 2015). According to his celebrated *interventionist theory of causation*, interventions uncover causal (explanatory) relevance relations. Since Woodward nicely captures certain intuitive aspects of empirical research, many have adopted this approach to capture how scientists explain phenomena. Woodwardian interventions thus feature prominently in much of contemporary philosophy of science, specifically in mechanistic and even reductionist accounts of scientific explanation (e.g. Silva et al. 2013; Craver & Bechtel 2007; Craver 2007b; Leuridan 2012).

Yet, a note of caution is in order. First, although mechanistic explanations and interventionism share a common origin, importing interventionist manipulations into contemporary mechanistic accounts of scientific explanation raises a number of puzzles. These are mostly due to the problem of distinguishing between two different dependence relations: *causal* and *constitutive* relevance. Basically, the point is that constituents (parts) and mechanisms (wholes) stand in a (constitutive) non-causal dependence relation. Thus, they cannot be subjected to interventionist cause-effect analyses. For illustration consider the simple phenomenon of a clock showing the time. The clock's showing the time is implemented by a mechanism that moves around cogwheels inside the clock as well as hands on the face of the clock. The cogwheels and the hands are *component parts* of the clock. Both their movements are *constitutively* relevant to the clock's showing the time. The relation between them, by contrast is one of cause and effect: the motion of the cogwheels on the inside *causes* the motion of the hands on the clock's face. But the motion of the cogwheels or the motion of the clock's hands does not cause the clock to show the time—they constitute it. Mechanistic explanations laudably capture this difference between causal and constitutive dependence relations. Now the puzzle is this. Although causal and constitutive relations are clearly different in character, they are both *dependence* relations. Both the clocks's showing the time (T) and the motion of the hands on the clock's face (H) depend (in some sense) on the motion of the cogwheels (C). For instance, if we put a magnet right next to the clock (I) to interfere with C we will in turn interfere with both H (C's effect) and T (the whole C is a part of). This is exploited by the mechanist who suggests using interventions to uncover both causal and constitutive relations. And indeed this suggestion seems empirically realistic given what we see in scientific practice. However, using interventions to uncover constitutive relations is not compatible with the interventionist view; for wholes (like T) non-causally depend on their constitutive parts (like C).

Second, even if we can modify interventionism such that we can use it to assess constitutive relations as well, problems remain for contemporary "mechanisms and interventions"-style philosophers of science. Most pressingly, there are in fact many different ways in which scientists manipulate when they try to figure out how something comes about; and intervening on a (cause-) variable to elicit a change in another (effect-) variable is only one method among many. The focus on Woodwardian interventions in contemporary philosophy of science therefore is rather short-sighted. If we want to give an account of scientific explanations, especially for heterogenous special sciences such as cognitive neuroscience, we need to pay attention to a wider variety of experimental and inferential strategies employed by practicing scientists.

By carefully examining different manipulative strategies used in cognitive neuroscience, it becomes clear that only a very small subset of empirical research actually employs Woodwardian interventions. And those that do, often combine Woodwardian interventions with what I call *mere interactions*. Mere interactions are manipulations that are not illuminating *qua* their effect but *qua* the information that their application makes available to us. If, for instance, we stain a sample of neural tissue the interesting experimental result is not *that* the tissue is stained after applying certain chemicals to it—we know it normally will be. This is why we use a staining technique in the first place. The result we are interested in is which neurons project to one another. Although they did so all along, independent of the staining manipulation, it is the staining that reveals this information.

Mere interactions are best understood as tools that allow for the assessment of certain features that are otherwise unobservable. They can be conceptualized as putting something under a metaphorical *magnifying glass* or fixing background conditions to enable scientific observation. While mere interactions may seem somewhat weaker than Woodwardian interventions, it is important to emphasize that they are in no sense less valuable for empirical practice. In fact, mere interactions might be a precondition for carrying out genuine Woodwardian interventions. Or they may be employed to simulate Woodward-style interventions where genuine interventions are unavailable for ethical, practical, or other reasons. In such cases, sets of mere interactions are interpreted *as if* there had been Woodwardian interventions, I speak of *pseudo-interventions*.

Experiments not only differ with respect to what manipulations (mere interactions, pseudo-interventions, or Woodwardian interventions) they employ. Different experimental strategies also permit different inferences with respect to (i) (causal) dependence relationships, (ii) features of (componential or temporal) organization, or (iii) both. All of this information, often integrated across different studies, contributes to explanations of specific phenomena or processes which typically cite some sort of generalization of an observed regularity. Understanding how scientists come up with such explanations takes careful consideration of the *experimental designs* and the *materials and methods* scientists employ. Pushing forward into this direction is the main goal of this book. In the end, I will present a catalog of experiments classifying different empirical studies with respect to which manipulations they employ and which research questions they can answer.

1.3 "Materials & Methods" of This Book

This book engages with contemporary debates in the philosophy of science and introduces a project in philosophy of experimentation. It aims to draw philosophers' attention to what precisely scientists do in their laboratories, and to demonstrate how understanding empirical research illuminates philosophical theorizing. At the same time, practicing scientists may find it a helpful introduction to current issues in philosophy of science. As I tailor my discussions mostly to cases from cognitive neuroscience, this book may also be considered a book in philosophy of neuroscience, or perhaps philosophy of cognitive science.

Cognitive neuroscience is a relatively young special science, and an 'up and coming' branch of cognitive science. It studies a wide range of *cognitive* or *mental* phenomena employing a large pool of different methodologies.[2] The common conception among cognitive neuroscientists is that cognitive processes are somehow *grounded* in, *implemented*, or *realized* by neural processes in the brain. Thus, cognitive neuroscience faces some peculiar challenges about relating different *levels* or *domains* of investigation, such as psychological and neurophysiological. Relating mental and physical is a challenge familiar from philosophy of mind. However, this is not the place to delve into century-old debates about mind-brain relations. While my discussions do connect to these traditional debates in philosophy of mind, my focus is on scientific methodology, not metaphysics.

---

2 I shall use the notions "cognitive", "psychological", and "mental" pretty much interchangeably throughout this book. Though these terms do have slightly different connotations, "cognition" has largely become the post-behaviorist term for just about anything the mind does (cf. Miller 2003, p. 142), viz. what has traditionally been studied by psychologists.

To learn about the principle logic behind scientific experiments, cognitive neuroscience is an ideal place to start. For one thing, the tools and phenomena are so diverse that insights from this field are likely to be applicable to many other special sciences. For another, the particular interlevel character of many experiments bears significant challenges for philosophy of science. Learning how we can handle these in cognitive neuroscience will likely inform philosophy of science more generally.

As said, contemporary philosophy of science is dominated by debates on *mechanistic explanations* (e.g. Andersen 2014; Bechtel & Abrahamsen 2008; Craver 2007b,a; Craver & Darden 2013; Craver & Tabery 2015; Fazekas & Kertész 2011; Franklin-Hall 2016b; Gebharter 2014; Glennan 1996, 2002, 2010b,a, 2013; Illari et al. 2011; Kästner 2015, 2016; Kuorikoski & Ylikoski 2013; Machamer et al. 2000; Woodward 2013) and *interventions* (e.g. Baumgartner 2013; Gebharter 2016; Hoffmann-Kolss 2014; List & Menzies 2009; Kästner 2011, 2017; Shapiro 2010; Woodward 2003, 2008a, 2015).[3] Both theories are—to a certain extent—rooted in seventeenth century mechanistic philosophy, both present significant progress over logical positivist views, and both capture some important aspects of scientific practice. They form the core of contemporary "mechanisms and interventions"-style philosophy of science.

This is where my story begins. Indeed, mechanisms and interventions do some very good work for understanding empirical research, but they face several problems—problems that philosophers cannot solve from their (metaphysical) armchairs. But closely examining scientific practice can teach us how to do it. Once we examine different experiments and their materials and methods—if we engage in philosophy of experimentation that is—we will discover that there are different modes of scientific investigation, including passive observation and different kinds of manipulations as well as different combinations and interpretations of these manipulations. This gives us the information we are missing if we appeal to mechanisms and interventions only and thus helps us gain a better understanding of how scientists explain (not only) cognitive phenomena.

---

3 While mechanistic explanations are currently a "hot topic" (Dupré 2013) it is subject to debate whether all explanation in science is actually mechanistic in character. This is especially discussed in the context of *dynamical systems* (see chapter 3).

1.4 Outline & How to Read This Book

This book is a classical *three-act play* in structure; chapters 5 and 10 offer brief intermediate summaries. The play starts out by introducing the two protagonists (mechanistic explanations and interventionism) and their mission (to give an account of how scientists explain cognitive phenomena) in part I. Chapter 2 outlines the challenge. It sketches how the new field of *cognitive neuroscience* emerged. Scientists studying mental, or as they are nowadays called *cognitive*, phenomena have become increasingly interested in the brain. The development of powerful neural imaging techniques made it possible to study processes in the brain alongside cognitive ones. This is where a challenge arises for contemporary philosophy of science: if we want to understand how scientific explanation works in cognitive neuroscience we need to develop an account of how scientists explain cognitive phenomena by referring to processes in the brain as (part of) the explanans.

Over the last thirty years or so, philosophy of science has gone beyond law-based views and focused on *causal processes* and *mechanisms* instead. Chapter 3 describes the genesis of a modern mechanistic view and introduces Carl Craver's (2007b) seminal account of mechanistic explanations. Chapter 4 introduces James Woodward's (2003; 2008a; 2015) interventionist account of causation, *interventionism* for short. Like Craver, Woodward was inspired by early mechanistic theories. And like Craver's, Woodward's view captures some important (though different) features of scientific practice, especially of scientific practice as we find it in special sciences such as cognitive neuroscience. Both accounts have thus been celebrated for overcoming challenges of traditional law-based views while being particularly empirically adequate.

Yet, there are problems. We will encounter these in part II. Although typically treated as a 'package deal' mechanistic explanations and interventionism are not too easily married. As it stands, interventionism cannot handle *interlevel experiments* that are so prominent in cognitive neuroscience, viz. experiments linking cognitive phenomena to neurophysiological processes. But without appealing to interventionist concepts of *manipulability*, *relevance*, and *intervention*, mechanistic explanations do not get off the ground either. This is the topic of chapter 6. The main problem is that mechanistic constitution is intended as some kind of interlevel part-whole relation. Although the notion "constitution" is not spelled out very precisely, it is quite clear that mechanists are after some sort of *non-causal dependence* relation here. Interventionists, by contrast, are concerned with *causal* dependence relations only and explicitly exclude non-causal dependence relations from their analysis. Chapter 7 will illustrate that no matter what the exact metaphysics of mechanistic constitution will be, interventionists must find a way to deal with non-causal dependence—at least if mechanistic explanations are to import core interventionist ideas. As chapter 8 will show, however, it is no easy task to distinguish between causal and constitutive relations based on Woodward-style interventions alone—and that is even if we neglect interventionism's problems with non-causal dependence. Besides, it seems that there is much more to empirical practice than Woodward-stlye interventions anyway. And this is true, as chapter 9 will argue, even if we conceive of the interventionist project as just giving an account of how scientists draw causal inferences rather than reason about dependence relations more generally. At the end of part II, the situation seems rather bleak. Mechanistic explanations and interventionism do not seem to go together after all; and individually, neither seems up the task of giving a plausible account of how scientists explain cognitive phenomena.

Finally, in part III, the problems will be resolved in two steps. First, I will offer a new reading of interventions in chapter 11. Essentially, the idea is to conceive of interventions as not necessarily uncovering causal but *general* dependence relations. Adopting this revision—which I call difference making interventionism (DMI)—will finally allow interventionism and mechanistic explanations to be married as is typically suggested. However, this does not suffice for them to meet their challenge yet. In order to tell a plausible story of how scientists explain cognitive phenomena (even just those that are mechanistically explicable), more is needed. Second, chapter 12 introduces the distinction between two types of manipulations: Woodwardian interventions and *mere interactions*. As I see them, "mere interactions" are an umbrella term under which many different types of manipulations find shelter, none of which conform to Woodward's definitions of interventions. Some of these can still be given a Woodwardian interpretation, though; I shall call these *pseudo-interventions*. Once mere interactions and pseudo-interventions are taken into account, a more complete picture of how scientists explain cognitive phenomena can be drawn. This does not only mean contemporary philosophy of science gains empirical adequacy, it also fills the gap that mechanistic explanations and interventionism have left open. For instance, mere interactions illuminate how scientists identify potential components before testing for their (causal or constitutive) relevance. Realizing that scientists use a variety of methods to study different features of a single phenomenon or system highlights some interesting ideas as to how to conceive of interlevel relations in the context of scientific explanations. In a brief excursus, chapter 13 will sketch a *perspectival view* of levels which illuminates how some of the remaining issues connected to interlevel (mechanistic) explanations might be addressed. Finally, chapter 14 illustrates how different modes of scientific investigation, particularly different manipulative strategies (Woodwardian interventions, mere interactions, pseudo-interventions, or combinations thereof), are systematically employed in different kinds of experiments to answer different research questions.

The upshot is that supplemented with mere interactions, mechanistic explanations and a modified version of interventionism will be able to accomplish their mission, after all. Or, at the very least, they will get us a good deal closer towards understanding how wards understanding how scientists explain (cognitive) phenomena.

Part I: **Stage Setting**

"Almost all cognitive scientists are convinced that in some fundamental sense the mind just is the brain [...]. Cognitive scientists are increasingly coming to the view that cognitive science has to be bottom-up as well as top-down. Our theories of what the mind does have to co-evolve with our theories of how the brain works."

—Bermúdez (2010)

2 Braining Up Psychology

Summary

Traditionally, investigating mental phenomena has been the domain of psychology—an experimental science that itself only emerged in the late 19th century. The current chapter describes how the interest of scientists studying mental phenomena has shifted towards the brain in the past 50 years or so. As a result, the field of *cognitive neuroscience* emerged. Today, cognitive neuroscience is a highly specialized, yet interdisciplinary, research area that employs a range of different methodologies and a plurality of descriptions to study a wide range of phenomena. *Interlevel experiments* have been particularly successful in advancing our knowledge of both mental phenomena and their biological bases. Hence, this is the field to look at if we want to understand how scientists explain cognitive phenomena.

2.1 From Early Psychology to Cognitive Science

Historically, mental phenomena, processes, or capacities have been studied by philosophers.[1] Only in the late 19th century, an expressed empiricism in philosophy together with rising natural sciences led to the emergence of psychology as an experimental science of mind. Taking an experimental rather than a theoretical course, psychology became established as a science in its own right. Early experimental work includes Ebbinghaus' research on learning and memory in the mid 1880s, Louis Émile Javal's research on eye-movements during reading (around 1887) and Hermann Helmholtz's work (dated between 1850 and 1870) on measuring the speed of nerve signal transmission (cf. Hoffmann & Deffenbacher 1992). Wilhelm Wundt and his student Edward Titchener developed the foundations for a comparative study of mental processes by carrying out systematic measurements and experimental manipulations (cf. Wundt 1874, 1907).[2] Paul Broca's (1861) and Carl Wernicke's (1874) seminal work on *aphasias* also fell in this era. But its significant effect on psychology only became apparent in the mid 20th century.[3]

---

1 I shall use the notions "phenomena", "processes", and "capacities" in a non-technical way. For the purposes of my discussion I take them to be pretty much interchangeable. For more on the notion of phenomena see Feest (2016).

2 Not all psychologist at the time agreed on this physiological course, though. William James, for instance, emphasized experiential consciousness and its practical significance, or function, and the use of introspection.

3 I will discuss the Broca's and Wernicke's patients in more detail below.

DOI 10.1515/9783110530940-002

Since its establishment as an empirical science of the mental, investigating phenomena such as learning, memory, language processing, (object) recognition, (visual) perception, reasoning, inferencing, abstraction and generalization, planning, problem-solving, decision making, etc. has been the domain of psychology. However, after the rise of *behaviorism* between 1892 and 1912—especially in North America—inquiry into the mind was gradually replaced with inquiry into behavior. The perhaps most famous psychologists of this time were Burrhus Frederic Skinner and John Broadus Watson. Watson characterized psychology as follows:

> Psychology, as the behaviorist views it, is a purely objective, experimental branch of natural science which needs introspection as little as do the sciences of chemistry and physics. It is granted that the behavior of animals can be investigated without appeal to consciousness. (Watson 1913, p. 176)

It took a good three decades for psychology to recover from this neglect of the mental. This is not the place to go into the details of the so-called *cognitive revolution* of the 1950s (but see Bechtel et al. 1999; Kästner 2009). However, to see the long way that contemporary research has come let me briefly mention some key events.

Basically, two major developments came together to overcome the behaviorist doctrine. First, scientists recognized that there must be more to the processes in our heads than simply mapping behavioral responses to perceived stimuli (e.g. Tolman 1948). Research into the precise ways in which humans take in, internally process and react to information was facilitated by World War II. For a better understanding of human attention, learning, and performance was vital to train soldiers. Second, research in computation, information theory, and the advent of the computer (Turing 1936; von Neumann 1993 [1945]), again facilitated by the war, led scientists to draw parallels between computing machines and mental processing. Thus a new branch of psychology emerged that essentially conceptualized humans as *information processing machines*. Ulric Neisser (1967) named this field *cognitive psychology*.

Once behaviorism was overcome and the mind was put back into the head, it was quite an uncontroversial assumption that the brain must be the physical locus of mental—or, as it was now called *cognitive* (see Miller 2003)—processing. Indeed, researchers had not only discovered that computers can be used to model mental processes (Newell et al. 1958), but also that biological neurons share important properties with the electrical circuits in computing machines that follow the rules of logic (McCulloch & Pitts 1943; Piccinini 2004). This had researchers in the developing field of *cognitive science* realize that logic, computation and the brain are "all of a piece" (Boden 2007, p. 4).

2.2 The Rediscovery of the Brain

The suggestion that the brain is the seat of mental processes was hardly new; after all, it had already been made by Ancient Greek philosophers. Yet, and although the claim that neural computations explain mental activity was relatively well accepted back in the 1940s and 1950s, cognitive scientists' work at the time was relatively far removed from that of neurophysiologists. After all, the questions that interested cognitive scientists, particularly cognitive psychologists, were quite different from what neurophysiologists could investigate in their labs. Moreover, the *computer metaphor* made neuroscience somewhat uninteresting for psychologists: just like software engineers do not need to bother with hardware, there seemed to be no need for psychologists to bother with neurophysiology.

In the following years, the computer metaphor was increasingly challenged and computer scientists shifted their focus from symbol processing machines to connectionist networks for modeling cognitive processes (e.g. Kästner & Walter 2013). At the same time, neuroanatomists discovered more and more systematic connections between function and anatomy. For instance, Ungerleider and colleagues conducted a series of studies in monkeys which led them to conclude that visuo-spatial processing is supported by two functionally specialized cortical systems: a ventral "what" and a dorsal "where" or "how" stream extending from occipital to temporal and parietal areas, respectively (Ungerlieder & Haxby 1994; Haxby et al. 1994; Ungerleider & Mishkin 1982; Goodale & Milner 1992).

Also studying the visual system, Hubel and Wiesel (1959) discovered that specialized cells in cats' primary visual cortex selectively respond to stimuli in different sizes, positions, shapes, and orientations and that they are arranged in a columnar fashion. Looking at the visual system as a whole, Felleman and Van Essen (1991) identified 32 functionally different brain areas responsible for visual processing and identified more than 300 connections between them. Although visual processing has probably been studied in most detail, scientists have also been interested in other cognitive processes and their neural basis. A case in point are Broca's and Wernicke's observations in their aphasic patients mentioned above.

Aphasia is typically caused by a traumatic brain injury (e.g. due to accident, stroke, or encephalitis). Broca's patient Leborgne (better known as "Tan"; because "tan" was the only word he could speak) suffered from loss of speech while his language comprehension as well as other mental functioning remained fully intact. Patients with *Broca's aphasia*, viz. Leborgne's condition, typically suffer from articulatory deficits and agrammatism; they have difficulties discriminating syllables while auditory comprehension is preserved (e.g. Blumstein et al. 1977). After Leborgne died, Broca conducted an autopsy and found that his patient had a brain lesion in the left inferior frontal gyrus (IFG) (cf. Broca 1861)—the specific portion lesioned in Leborgne corresponded to Brodmann's areas (BA) 44 and 45 and has subsequently become known as *Broca's area*.

Complementing Broca's observations in Leborgne, Carl Wernicke (1874) reported about a patient who displayed the reverse pattern of impairments: though perfectly fluent in language production, his speech was non-sentential and he clearly displayed severe deficits in language comprehension despite intact syllable discrimination (Ward 2006)—a condition known as *Wernicke's aphasia*. An autopsy revealed damages in the patient's left posterior superior temporal gyrus (STG) corresponding to the posterior part of BA22 and known as *Wernicke's area*.[4] This clearly established that Broca's area is not the only part of the brain relevant for language processing. More than that, the clinical observations in Broca's and Wernicke's patients suggested a *double dissociation* between language production and language comprehension where damage to Broca's area (BA44/45) affects speech motor output whereas damage to Wenicke's area (posterior BA22) affects speech motor comprehension (Geschwind 1967).[5]

Although Broca's and Wernicke's reports are much older than e.g. Ungeleider and Mishkin's or Felleman and Van Essen's research, they too have been discussed vigorously in the late 20th century. They make it just as evident that specific brain areas are specialized for certain cognitive functions as do the more recent studies on visual processing.[6] This very discovery—that there is functional specialization within the brain and that anatomy seems, after all, to be highly relevant to how cognitive processing works—led to an increased collaboration between neuroscientists and psychologists.

The development of powerful functional imaging methodologies that allowed "watching the brain at work" further facilitated combined research into cognitive and neural processing.[7] Michael Gazzaniga, himself well-known for his studies in split-brain patients, is credited for naming this new field *cognitive neuroscience* in the late 1970s.

---

4 Note that neither Broca's nor Wernicke's patients were actually as "clean" a case as I present them here. But for current purposes, this textbook-like version shall suffice.

5 For details on dissociations see p. 41.

6 In fact, linguistic processing may be analyzed in terms of "what" and "where"/"how" streams analogous to the streams identified for visuo-spatial processing; I will get to this in chapter 2.3.2. Since basically the same functional organization is observed in language processing and visual processing, respectively, this may indicate a general organizational principle of brain (for details on this see Kästner 2010b).

7 See Senior et al. (2006) for an overview of some of the standard methodologies used in cognitive neuroscience, including discussions on their vices and virtues.

2.3 Cognitive Neuroscience Today

Today, the dominating assumption shared among cognitive scientists is that phenomena such as vision, learning, memory, language, or reasoning are somehow *brought about* or *produced* by the mind, which, in turn, is typically understood as some kind of information processing system that (at least in humans) is biologically grounded in or *realized by* the brain. José Bermúdez puts this clearly:

> Almost all cognitive scientists are convinced that in some fundamental sense the mind just is the brain, so everything that happens in the mind is happening in the brain. (Bermúdez 2010, p. 6)

Hence, it seems that if we want to understand the mind we must understand not only cognitive (or psychological) capacities as we observe them but also the internal workings of the brain on which they are based. In order to achieve this, many argue, "cognitive science has to be bottom-up as well as top-down. Our theories of what the mind does have to *co-evolve* with our theories of how the brain works." (Bermúdez 2010, p. 87) This is why cognitive neuroscience has become an indispensable part of contemporary cognitive science.

While brains and neurons can generally be studied without recourse to the psychological domain (e.g. by molecular biologists and cellular neuroscientists) and psychological phenomena without recourse to neural processes (e.g. by cognitive or behavioral psychologists), there now is an alternative to such *same-level approaches*: truly in the spirit of co-evolving mind-brain theories, studying a psychological phenomenon can be combined with the study of the brain processes that accompany it.[8] Ultimately, the aim of such an *interlevel approach* is to associate specific neural processes in the brain with phenomena observed in the psychological domain and thereby localize the phenomenon's physical substrate (e.g. in a specific brain area). This is the domain of contemporary cognitive neuroscience.[9]

Before ending this chapter, let me briefly outline two success stories of cognitive neuroscience: how it progressed our understanding of memory and language processing, respectively. While they both are paradigmatically cognitive phenomena, they are actually quite different and place different constraints on the methodologies that can be used to study them. For instance, memory can relatively easily

---

[8] For the time being, I shall use the term "level" in a non-technical way. It merely expresses the common assumption that cognitive processes are somehow 'higher than' and 'in a different domain from' neural ones.

[9] Tough decomposition and localization are prominent in cognitive neuroscience, recently other approaches, such as *dynamical modeling* become more and more important. For now my discussion will focus more on localization and decomposition strategies, though.

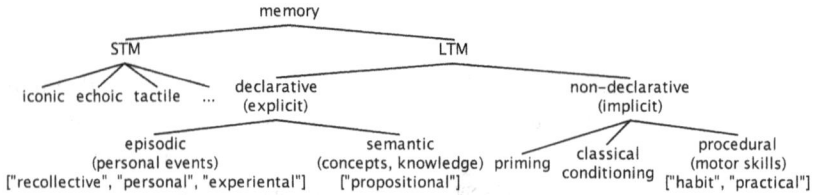

Fig. 2.1: A taxonomy of memory; schematic overview of memory types as they are typically classified in contemporary psychology (e.g. Squire 2004; Goldstein 2004; Gazzaniga et al. 2008).

be studied in model organisms while language cannot. Yet, there are some important commonalities in what strategies researchers employ when trying to explain memory and language processing, respectively. The remainder of this section will illustrate this.

2.3.1 Progress in Memory Research

Memory used to be seen as a single, unified cognitive capacity. Yet, "memory" is a remarkably rich concept: I remember that I wanted to be a doctor when I was little, remember how to ride a bicycle and how to play Go, and I also remember to brush my teeth every night. I remember the smell of peppermint, the date of my sister's birthday and what I did the night Barack Obama was elected president. The wide range of examples demonstrates what already Wittgenstein realized and contemporary psychologists and neuroscientist will not hesitate to confirm: "Many different things happen when we remember." (Wittgenstein 1974, p. 181)

In recent years, scientists have come to sort different instances of memory into classes; see figure 2.1 for an overview. Particularly studies with lesion patients and observed dissociations between memory impairments have advanced the scientific distinctions between different types of memory and their neurobiological implementation. Since this is not the place for a full historical survey, I shall just pick a few well-known milestones from memory research for illustrative purposes.[10]

2.3.1.1 Patient H.M.
The perhaps most famous patient among memory researchers is Henry Gustav Molaison, better known as H.M. (1926-2008). Suffering from intractable epilepsy,

---

10 For a fuller and quite easily accessible overview on the history of memory research, readers may consult Eichenbaum & Cohen (2004).

he underwent neurosurgery in 1953. During the operation, parts of H.M.'s medial temporal lobes (MTL) of both hemispheres have been removed whereby H.M. lost approximately two thirds of his hippocampus, his amygdala, and parahippocampal gyrus. Although successful in controlling his epilepsy, the surgery had dramatic effects on H.M.'s memory function. Most remarkably, H.M. suffered from severe anterograde amnesia; that is he was no longer able to form new long-term memories, while his working memory as well as some motor-learning skills remained intact (e.g. Milner et al. 1968).

Without digging into the details of how the current taxonomy of memory has unfolded throughout the history of neuroscience, I take it that the principle is quite clear: whenever it is possible for someone to recall some events (say, those that happened a long time ago) but not others (say, those that happened just a minute ago), it cannot be the case that both types of memories rely on the very same neurological processes. On the contrary: such a pattern of impairments indicates a *dissociation* between the affected and non-affected cognitive functions, e.g. long-term memory (LTM) and short-term memory (STM).[11]

Studying lesion patients such as H.M. (no matter whether the cause of the patient's lesion was surgery, an accident, or something else) is particularly valuable when trying to understand cognitive functions and how they are implemented by the brain: they can easily demonstrate features of human cognitive architecture that scientists may not have thought of otherwise—such that memory decomposes into various distinct subtypes—or falsify established models by providing evidence for a dissociation incompatible with those models.

Given the patterns of memory impairments H.M. suffered from, he was diagnosed with a deficit in transferring facts from STM to LTM. Since this deficit occurred as a result of his surgery, while other cognitive functions remained largely intact, the missing function was attributed to the removed brain tissue: the portions that had been removed from H.M.'s brain must be the structures normally hosting the affected functions. And indeed, H.M. was not the only patient with this kind of impairment after hippocampal resection. In their classic paper, Scoville and Milner (1957) report about ten cases all pointing to the importance of the hippocampal formation for normal memory function, particularly the formation of recent (anterograde) memories.

However, neurophysiologists of the time hesitated to associate hippocampus with learning and memory formation. Yet, they studied it quite intensively because for one thing it was associated with epilepsy, for another it was a model of neural circuitry with a relatively simple organization—which was rather easily accessi-

---

11 See chapter 2.5 for more details on dissociations.

ble in model organisms such as rats—and hence very valuable for experimental practices (cf. Craver 2007b, ch. 7).

2.3.1.2 Hippocampal Long-Term Potentiation

In 1966, Terje Lømo in Oslo first discovered hippocampal long-term potentiation (LTP) in rabbits (Lømo 2003). LTP is a long-lasting enhancement of synaptic connections between neurons that results from repeated synchronous electrical stimulation of the axonal pathway which leads to longterm increases in the size of postsynaptic potentials. LTP thus augments signal transmission between the affected neural units. In fact, LTP is just one out of several phenomena underlying synaptic plasticity, viz. the ability of terminal nerve-endings to modify the strength of their connections to other neurons, but we shall not dwell into more details at this point.

Hippocampal LTP has been studied extensively, especially in model organisms such as squids and rats. During the early days of its investigation, scientists have primarily focused on how the neural mechanism itself works rather than on how it may implement complex memory functions. But the connections to higher-level cognitive functions were soon to be made.

The discovery of hippocampal LTP converged with Donald Hebb's (1949) theory of learning. In his "The Organization of Behavior" Hebb had theorized that connections between neurons are strengthened if these (the neurons) become simultaneously activated; true to his well-known motto "What fires together, wires together." The LTP mechanism discovered by Lømo seemed to be doing exactly what Hebb had been looking for: cell assemblies undergo long-term morphological changes such as development and growth of synaptic connections in response to repeated stimulation; neural activity thus can modulate synaptic connectivity.[12]

In subsequent years, hippocampal LTP has become widely considered the major cellular mechanism that supports consolidation (the process by which memory becomes encoded) though modification of synaptic connections.[13] More recently, researchers have also been manipulating specific types of neuroreceptors, such as N-methyl-D-aspartate (NMDA) receptors (e.g. Baddeley et al. 2009) as well as genetic factors (e.g. Kandel 2006) to see what role they play in memory formation and LTP. Quite prominently, hippocampal LTP has also come to feature in the contemporary debates on scientific explanation as it has been shown to play a

---

12 While LTP induces long-term changes and thus is held responsible for learning and LTM formation, STM is believed to be cortically realized by more temporary cortical activity.
13 See for example (Kaneld et al. 2000, ch. 62) for details on the LTP mechanism and how it supports learning and memory.

pivotal role in spatial memory (e.g. Craver & Darden 2001; Craver 2007b). I will come to this below.

Thus far I have been illustrating how our current understanding of memory has been advanced by reasoning about dissociations that links neurological to behavioral data, e.g., lesion data and data on cognitive deficits in clinical patients. I also sketched some neuroscientific research into the cellular and molecular mechanisms of hippocampal LTP. Yet, it remains unclear at this point *how* this mechanism supports memory; how, that is, hippocampal LTP is connected to memory processes. To be able to *explain* memory processes, however, this is precisely what is needed. How did scientists achieve to link cognitive phenomena with neural processes? To illustrate this, I will look at three different areas of memory research in turn: spatial memory, memory consolidation during sleep, and recognition memory.

2.3.1.3 Spatial Memory

The prime example from neuroscience discussed in the context of scientific explanations is spatial memory. Put very briefly, spatial memory is commonly understood as the part or type of memory that is responsible for keeping track of where we are, recording information about our environment, and spatial orientation. It is required, for instance, to navigate around familiar environments or find specific locations. LTM, STM as well as working memory (WM) are believed to contribute to spatial memory. In rats and mice, a typical paradigm assessing spatial memory is the Morris water maze—a tank filled with milky water that contains a small platform somewhere just beneath the surface. If a mouse or rat is put in the maze for the first time, it will swim around until it locates the platform and, since rats dislike water, sit on the platform as soon as it discovers it. If the animal is put back into the water, and if its spatial memory is intact, it will usually not hesitate to and swim straight back to the platform.[14]

According to the received view, the performance of such tasks requiring spatial memory crucially depends on neural activations in the hippocampus. More specifically, it is believed that hippocampus encodes a *spatial map*; that is, it summarizes and stores information about the agent's environment. These spatial maps enable us to locate ourselves when we navigate a city and that allow us to mentally walk along a familiar road, etc. The mechanism responsible for encoding and retrieval of spatial maps in hippocampus is widely believed to be LTP. LTP, in turn, crucially depends on the activation of certain types of neuroreceptors, viz.

---

14 Studies like these have been conducted in both mice and rats; both are prominently used as model organisms in memory research. I shall gloss over the differences between them for now.

NMDA receptors (for details see e.g. Craver & Darden 2001); figure 2.2 illustrates this implementational hierarchy.

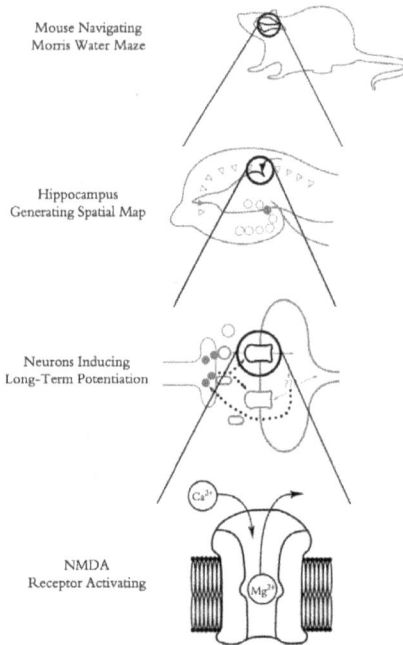

Mouse Navigating
Morris Water Maze

Hippocampus
Generating Spatial Map

Neurons Inducing
Long-Term Potentiation

NMDA
Receptor Activating

Fig. 2.2: Levels of spatial memory. From the animal's navigation behavior down to NMDA receptor activation. Adopted from Craver 2007b: Fig. 5.1 (p. 166). By permission of Oxford University Press.

Now how did scientists come to this view? There was not a single crucial experiment, of course, but a series of successive experiments that helped illuminate the phenomenon of spatial memory and its neural basis. For current purposes I shall just focus on two quite well-known studies that are illustrative: a lesion study in rats as well as a functional imaging study in humans. What is important about these studies is that they exemplify a paradigmatic strategy in cognitive neuroscience: they systematically relate processes in two different domains by manipulating in one and assessing the effect of the manipulation in another. In contemporary philosophy of science, these kinds of studies have become known as *interlevel* or *cross-level* studies. Depending on where manipulations are applied and effects are observed, and assuming some kind of hierarchy between different domains—or, as they are typically called, *levels*—interlevel studies can be either bottom-up or top-down.[15]

Lesion studies, such as the case of H.M. mentioned above, are an instance of what we may call *bottom-up studies*. That is, studies that look at what happens to higher-level (cognitive) processes once we interfere with lower-level (neural) processes that we believe the higher-level (cognitive) processes are based on. A prototypical example of a bottom-up study in spatial memory research is Morris et al.'s (1982) lesion study in rats: animals were trained on a Morris water maze task; subsequently, animals assigned to the experimental condition underwent hippocampal lesioning while those in the control condition did not. As a result, Morris et al. report that control animals performed significantly better when tested on the water maze than did the experimental group. This indicates that spatial memory performance, assessed through performance in the water maze task, is negatively affected in animals with hippocampal lesions. Importantly, the observed impairment is relatively isolated; as Morris et al. report, it "can be dissociated from correlated motor, motivational and reinforcement aspects of the procedure." (Morris et al. 1982, p. 681) Researchers are thus led to conclude, that hippocampus plays a vital role in spatial memory processes. Analogous results in the experimental group can be produced by leaving hippocampus intact but interfering with NMDA-receptor functioning—e.g. pharmacologically or genetically—and thus blocking NMDA-receptor dependent LTP induction (e.g. Tsien et al. 1996).

Bottom-up studies can generally be supplemented by complementary *top-down studies*—and, of course, vice versa. Top-down studies investigate what happens to lower-level (neural) processes once we interfere with higher-level (cognitive) processes that we believe the lower-level (neural) processes to implement. Top-down studies complementing the lesion studies with rats navigating the Morris water maze thus are experiments that assess the hippocampal activations in rats while navigating a maze contrasting them with the hippocampal activations measured in the animals during rest. Assuming that both experiments tap into the same cognitive phenomenon and its neural substrate, one should expect that a significant increase in hippocampal activations is observed when the rats engage in a spatial navigation task rather than rest. In fact, such results have been reported prior to Morris et al.'s lesion study (cf. e.g. O'Keefe & Dostrovsky 1971; Olton et al. 1978).

---

15 I adopt this terminology from Craver (2007b). For more on Craver's account of mechanistic explanations see chapter 3. Note that there also is a different sense of "bottom-up" and "top-down" experiments that appeals to Marr's (1982) computational, algorithmic, and implementational levels (see Zednik & Jäkel 2014). I will work with a less technical sense of "levels" for now. But see chapter 13 for how I would like to see the notion understood.

Arguably, bottom-up studies in cognitive neuroscience are often preceded by top-down studies indicating a rough localization for the cognitive capacity in question. This is for two reasons: first, it is often easier to measure neural potentials than to remove neural tissue, and second, if one looks for the neural *substrate* of a cognitive phenomenon, the obvious way to do this is to induce the very phenomenon in question in healthy individuals and see what happens in the neural domain at the same time. Morris et al. probably would not have lesioned the rats' hippocampi if they had not already suspected, based on previous electrophysiological studies, that this would yield the anticipated effect. However, we may also think of cases where bottom-up studies precede top-down research: think of, e.g., a clinical patient with a focal brain lesion who displays a specific cognitive impairment. The discovery of his lesion may initiate subsequent top-down research into the neural substrate of the affected phenomenon. But the temporal order is not at issue here. For the time being, all I want to demonstrate is how interlevel studies are used in cognitive neuroscience.

Also note that there is an intimate link between dissociations and interlevel manipulations. On the one hand, it is important to delineate a cognitive phenomenon from related processes before investigating its neural basis. On the other, complementary interlevel studies can be used to confirm or reject hypothesized dissociations or indicate new ones.

Once we know that the hippocampus plays a central role for spatial memory in rats, one may wonder whether this translates to humans. That is, one may ask whether hippocampus is engaged in human spatial memory, too. However, neither single cell recordings from deep brain structures (which requires electrode implantation) nor well-controlled lesion studies (requiring deliberate brain damage) are easily available in humans—and if patients with focal brain lesions or implanted electrodes are available, they are mostly single cases and often suffer from severe additional clinical conditions. Luckily, though, cognitive neuroscientists can refer to a range of non-invasive neuroimaging techniques that measure brain activations in healthy individuals and thus enable scientists to draw inferences with respect to which brain structures are engaged in different tasks. The most well-known instances probably are electroencephalography (EEG), magnetoencephalography (MEG), and (functional) magnetic resonance imaging ((f)MRI). While EEG and MEG measure electrical and electro-magnetic potentials on the surface of the skull, respectively, fMRI measures cerebral blood flow which is used as a proxy for neural activations. Note that all these are indirect measures only. Moreover, temporal and/or spatial resolution—depending on methodology and target structure—tend to be problematic.[16] Yet, such technologies are the best tools we have; they enable researchers to study the brain of living humans without doing them any harm.

A well-known example of a neuroimaging study investigating spatial memory in humans is Maguire, Woollett and Spiers' (2006) structural MRI study in London taxi drivers. Their study revealed increased gray matter volume[17] in mid-posterior hippocampi and reduced gray matter volume in anterior hippocampi in London taxi drivers compared to London bus drivers. Drivers in both groups were matched for factors such as driving experience and stress-levels, but differed with respect to the routes they take: taxi drivers have to freely navigate requiring complex spatial representations of the city while bus drivers follow a constrained set of routes. This result suggests that spatial knowledge can indeed be associated with the specific pattern of hippocampal gray matter volume observed in London taxi drivers. Put slightly differently, and idealizing away from methodological limitations, taxi drivers' extraordinary navigation skills can be associated with an increased number of neural cell bodies in posterior hippocampi (as well as the decreased number of neural cell bodies in anterior hippocampi). Therefore, since we know that such structural changes are indicative of functional characteristics, we can infer that spatial memory in humans is supported by certain hippocampal structures, too.[18]

In the current section, I have been focusing on interlevel studies investigating spatial memory. I do not contend this brief illustration to be anywhere near a full story of how spatial memory is neurally implemented in the brain or how the neural structures supporting it have been discovered. But I do take it to show a glimpse of how empirical research in cognitive neuroscience contributes to scientific explanations of cognitive phenomena in terms of neural process. The reader may object the case has been cherry picked. After all, spatial memory is *the* paradigm example in contemporary debates about scientific explanations. As the following sections will demonstrate, it is not cherry-picked, however. In fact, as I will continue to

---

16 This is not the place to give an overview on the different available neuroimaging techniques and how exactly they work. For a comprehensive introduction to some standard methods the reader is referred to Senior et al. (2006).

17 Gray matter consists of neural cell bodies whereas white matter primarily consists of glial cells and myelinated axons, viz. fibers relevant for signal transmission. Therefore, increased gray matter volume indicates the presence of more neurons and thus more information processing units.

18 Of course, spatial navigation is not all that hippocampus does. Also, it is unlikely that all of hippocampus is engaged in spatial map generation. Maguire et al. (2006) speculate that posterior hippocampi host complex spatial representations enabling sophisticated navigation whereas anterior hippocampi seem to be more engaged in acquiring new visuo-spatial information—here again, we also see how reasoning about dissociations is intimately linked with interlevel research. A fuller discussion of these issues must be left for another occasion, however.

show below, we do find this very pattern of interlevel research not only in different areas of memory research but also in research targeting quite a different cognitive phenomenon, viz. language processing.

2.3.1.4 Consolidation & Plasticity

Although hippocampus is still believed to play a vital role in (at least some) memory processes, other brain areas as well as mechanisms beyond LTP are now believed to selectively contribute to various different memory processes. Rather than being in contradiction with the spatial memory research discussed above, these findings are best viewed as specifications and extensions. For instance, different types of plasticity-induced memory formation seem to be based on synchronous oscillations at specific frequencies. That is, it is not just synchronous activation inducing LTP that is important for consolidation but it is also important *at which frequencies* the neural networks get activated.

Generally, cortical oscillations can occur at different temporal and spatial scales. Which patterns of oscillatory activity arises in a network is dependent upon (i) its precise neural architecture and (ii) the initial conditions (baseline activity, noise, thresholds) of the network and its constituent units (e.g. Buzsaki & Draguhn 2004; Kirov et al. 2009). Alpha (α) activity (8–13Hz) is typically associated with resting states during wakefulness while beta (β) activity (13–25Hz) is normally found when the subject is either concentrated and alert or in rapid-eye-movement (REM) sleep. Gamma (γ) activity (>25Hz) can be further subdivided into low (25-50Hz) and high (50-140Hz) γ, though the exact nomenclature is not consistent across studies. Both gamma and theta (θ) (4–8Hz) are believed to play a "mechanistic role" (Düzel et al. 2010) in memory processes such as formation of representations, off-line maintenance, and retrieval. Theta activity is further found in infants and sleeping adults. In very deep sleep, slow delta (δ) waves (0.5-4Hz) are observed. Oscillatory activity at these different frequency bands is thought to mark different cortical processing modes and brain states (Thut & Miniussi 2009) where neighboring frequencies may reflect competing brain states (Buzsaki & Draguhn 2004). The basic idea behind these statements is that structures that oscillate in concert jointly engage in one kind of processing: oscillatory activities at a given frequency band are thus indicative of specific cognitive processes, e.g. subsequent stages of memory formation (Axmacher et al. 2006) or sensori-motor processing and executive functioning (Thut & Miniussi 2009). As different cognitive tasks place their demands on the system they may "compete" about the neural processing resources.

Generally, slow rhythms are associated with synchronous activity in large spatial domains whereas higher frequencies accompany faster more local events in the brain. Different types of oscillations can co-exist in the same and/or different brain structures and even reinforce or inhibit one another (e.g. Buzsaki & Draguhn 2004). Although it remains to be investigated how exactly neural oscillations work together to facilitate (human) cognition, rhythms at different frequencies have been found to relate to specific cognitive functions.

Without delving into too much detail at this point, let me just consider memory consolidation during sleep. As Buzsaki & Draguhn (2004) report, information learned during wakefulness is continuously replayed during sleep inducing functional long-term changes. Human early sleep is dominated by so-called slow wave sleep (SWS), i.e. sleep during which cortical oscillations mainly occur at frequencies below 1Hz. This phase of sleep has been shown to specifically benefit hippocampus-dependant declarative LTM. Human late sleep, on the other hand, is dominated by so-called rapid-eye-movement (REM) sleep and benefits procedural more than declarative memory encoding (Plihal & Born 2008).

Note that, again, the studies leading to these conclusions are saliently inter-level in nature: participants engage in a cognitive task (declarative or procedural learning) while their cortical oscillations are measured using EEG during subsequent sleep, i.e. when participants consolidate the experimental task. Similarly, experiments in model organisms can selectively prevent neural oscillations at specific frequency bands to observe whether this interferes with declarative and/or procedural memory formation. Such bottom-up studies serve to complement the top-down EEG study described. Note again, that the investigation into the neural substrates of declarative and procedural LTM encoding, builds on a (primarily) behavioral dissociation between the two.

2.3.1.5 Recognition Memory

With the previous example I have been focusing my attention on research into memory encoding. Let us now turn towards memory retrieval by looking at research into recognition memory. Retrieval from LTM can be examined within a dual-process framework that defines two component processes for recognition memory, viz. our ability to recognize people, objects, or events we have previously encountered, and to associate specific details with them: familiarity and recollection (e.g. Jacoby & Dallas 1981).

In *recollection*, recognition of a particular item is based on an association of several features of this item, often based on contextual or episodic information; it typically requires integration of different kinds of information (sensory, temporal, spatial, etc.). *Familiarity* based recognition relies on a vague feeling of déjà-vu; it is memory of the recognized item *per se*. Familiarity based recognition can build on conceptual knowledge and/or associations but typically does not require recourse to contextual cues. Though both familiarity and recollection often occur in concert, they can be dissociated in various ways.[19] Behaviorally, the two types of recognition can be assessed through different cognitive tasks. To assess recollection based recognition, participants can be asked to perform a relational recognition task. Relational recognition tasks require participants to rely on their memory of specific features of the remembered items. To assess familiarity based recognition, participants can be asked to perform fast item recognition tasks that require them to rely on the overall appearance of the remembered item rather than any specific features.

Searching for the neural basis of recognition memory, researchers have found that successful recognition is associated with enhanced hippocampal activity. For instance, Stark & Squire (2000) conducted a functional imaging study where participants were asked to perform control as well as recognition memory tasks while lying in an fMRI machine. Comparing recorded signals revealed that changing from control to recognition memory condition changed something about the neural processes in the brain, viz. it enhanced hippocampal activations.

Adding to this evidence, psychologists have reported about patients suffering from severe deficits in recognition memory while their other cognitive abilities— including other types of memory—have been relatively spared. Brain scans of these patients revealed focal damages to hippocampal regions (see e.g. Manns et al. 2003). These case studies complement the results reported in the imaging study: if recognition memory yields hippocampal activations, and if one concludes this is the case because hippocampus plays a key role in recognition memory, hippocampal lesions should be expected to negatively affect recognition memory performance.

However, this is not yet satisfying. Especially in light of the behavioral or psychological analysis of recognition memory as composed of recollection and familiarity based recognition, the question arises what exactly the role of hippocampus in recognition memory is. If both processes are supported by hippocampus, are the two actually distinct or do they rather reflect different aspects of the same process? And even if they are distincit, are they in fact based on the same neural substrate?

First off, there is evidence that recollection and familiarity based recognition are in fact distinct processes as they are differentially affected in ageing and amnesia: while amnesia disrupts recollection, ageing more severely affects familiarity based recognition (e.g. Rugg & Yonelinas 2003). Moreover, as Gruber and colleagues (2008) report, the two can be dissociated at the neural level: while

---

19 I will return to this issue in more detail below.

induced θ in distributed areas is observed to reflect recollection processes, induced γ in predominantly occipital areas seems to be indicative of associative/conceptual processing. Similarly, Guderian & Düzel (2005) associate recollection with induced θ-synchrony in a distributed (prefrontal, visual, and mediotemporal) network. Now given that recollection and familiarity based recognition in fact appear to be distinct processes, and given that hippocampus plays an important role in recognition memory, the natural question to ask is whether hippocampus supports familiarity, recollection, or both. Recent research in humans and rats has helped illuminate this question.

In addition to the separate assessments of recollection and familiarity based recognition through different tasks, it is possible to assess how much participants rely on recollection and familiarity, respectively, in a single recognition task. This methodology is known as receiver operating characteristic (ROC) (see e.g. Yonelinas & Parks 2007):

> In humans, episodic recognition memory is usually assessed by presenting a study list of items to a subject (for example a list of words appearing on a screen one at a time), and after a delay, presenting a longer list of items composed of the same items intermixed with an equal number of new items, also appearing one at a time. The probability of correct recognition of a study list item [...] is plotted as a function of the probability of incorrect recognition of a 'new' item [...] across confidence or bias levels, and the best fitting curve is defined to generate an [sic] ROC function [...]. (Sauvage 2010, p. 817)

Neglecting the precise details of the ROC function, the basic idea is that one can read off the contribution of recollection and familiarity based recognition, respectively, from the shape of the plotted ROC function. Now assuming this approach is valid, we can straightforwardly design an interlevel experiment assessing the role of hippocampus in recognition memory: in a manner analogous to the lesion studies described above, scientists could lesion different parts of hippocampus and test whether or not this affects recollection and/or familiarity based recognition abilities. If hippocampus selectively supports recollection, then patients with hippocampal lesions should perform recognition tasks in such a way that they rely more on familiarity, and vice-versa. Whether recollection and/or familiarity based recognition is used, and to what degree participants rely on them, will be evident from the respective ROC curves.

Since controlled lesion studies in humans are unavailable for ethical reasons, scientists have to refer to animal models for such studies. However, this comes with a challenge: the tasks administered to assess recognition abilities in humans cannot simply be used with rats. Instead, an analogous task must be found that is appropriate for the rat. Complications aside, if the translation of the cognitive task is successful and the animal model is appropriate, then it should be possible to translate the results from the animal study back to the human case.Using a *translational* ROC *approach* in rats, Fortin et al. (2004) found that rats with hippocampal lesions rely more on familiarity than recollection in recognition memory tasks compared to unlesioned control animals. This suggests that hippocampus cannot support familiarity but selectively supports recollection based recognition.[20]

For a complementary study in humans, we can, once again, refer to functional neuroimaging. In a high-resolution fMRI study, Suzuki et al. (2011) report increased activity in a cluster of voxels in right hippocampus that was positively correlated with recollection performance. Taken together, the studies just sketched speak to the conclusion that hippocampus not is involved in recognition processes but specifically supports recollection.While the human fMRI data leave open whether or not familiarity based recognition is also supported by hippocampus, the ROC data from rat studies clearly indicate that hippocampus cannot support familiarity based recognition; for animals with hippocampal lesions relied on familiarity when performing recognition tasks.

In summary, the three snippets of memory research I have been describing in the current section clearly demonstrate the importance of interlevel studies in cognitive neuroscience. Bottom-up and top-down experiments are systematically combined in empirical research to rule out competing hypotheses and thereby advance our understanding of how cognitive phenomena are grounded in our biological brains. Yet, critical readers may object that this view of the matter only is appropriate for phenomena that we can pin down to low-level neural processes, such as synaptic plasticity or, even more skeptically, that it is appropriate for memory only. As I will continue to show in the following section, neither is the case. We do find the same principal pattern of experiments and reasoning about experimental manipulations also in research that investigates quite different, though still paradigmatically cognitive, phenomena such as (sign) language processing.

2.3.2 Progress in Language Research

Language arguably penetrates much of human thinking and cognition; some even hold there cannot be thoughts without utilization of some language (e.g. Bermúdez 2003; Fodor 1975). Although I would not go as far as claiming this, language clearly prominently features in many cognitive tasks to enhance our cognition. Language is characteristically systematic and productive; we can, for instance, use its compositional symbols to communicate about things not present in our immediate

---

20 For a detailed review of ROC paradigms in rats see Sauvage (2010).

surroundings, to scaffold memory, to engage in coordinated activities, etc. (e.g. Rumelhart et al. 1986; Clark 1998).[21]

I already introduced Broca's and Wernicke's aphasic patients above; they probably presented the first milestones in research into language processing. Unlike memory, language processing is not available for study in model organisms. Thus, its study heavily relies on clinical patients as well as the use of neuroimaging techniques. Once these had become available, scientists studied language processing extensively; sophisticated models of cortical processing of both spoken and signed languages have since been developed.

I shall start my illustration by briefly sketching a recent model of cortical spoken language (SpL) processing. Since this builds almost exclusively on spoken language research, an interesting question remains unanswered: do the neural processes associated with with SpL processing truly characterize language processing *per se* or do they rather characterize language processing in the auditory domain only? Sign languages (SLs) offer a unique opportunity to answer this question as they are not spoken and heard but produced by the body and hands and seen by the eyes. Thus, studying SL can help us dissociate possible modality-specific processes in the brain from linguistic ones. I shall therefore, in section 2.3.2.2 turn towards recent views on cortical SL processing.

Research into both SpL and SL processing has markedly been driven by dissociative approaches as well as interlevel studies of the kind introduced above. Finally (in section 2.3.2.3), I shall turn to an ongoing cross-linguistic SL study and illustrate its inherently interlevel approach in detail. Even though this example may seem quite removed from the memory research I have been discussing above, the research strategies employed in both cases resemble one another in a crucial point: scientists utilize the same basic experimental paradigms, whether investigating higher cognitive phenomena, such as language, or phenomena they can (quite straightforwardly) track down to neural or even molecular mechanisms, such as memory. Indeed, the methods I sketch here are abundant all over cognitive neuroscience.

2.3.2.1 The Speaking Brain's "What" and "How"
Speech is sound produced by vocal organs of a speaker; it typically ranges from 80Hz (deep male voice) to 400Hz (child voice). The speech signal is a train of *phonemes*, viz. smallest perceptually distinct units of speech; it is parsed (i.e.

---

21 I do not contend these functions are exclusive to language. I take it, however, that language is a powerful cognitive capacity and a valuable tool that gets employed in various different circumstances.

analyzed) by the listener into words, which are subsequently parsed into sentences, assigned meanings, and given a (context dependent) interpretation. But how is this complex linguistic processing achieved by the brain?[22]

As Davis & Johnsrude (2003) suggest, an initial phonetic analysis of the speech signal may be localized predominantly in superior temporal regions. This lines up with Leech and colleagues' (2009) finding that on-line analysis of an input's specto-temporal structure (silent gaps, noise burst, formant transitions)—crucial for classification of speech sounds—recruits superior temporal sulcus (STS) and superior temporal gyrus (STG). Similarly, Fred Dick and colleagues (2011) report increased activity in right STS, and left superior anterior STG when expert violinists are listening to music as opposed to speech. Taken together, these findings support the view that superior temporal regions are crucial for the initial analysis of auditory speech signals on the one hand, but beg the question against their language specificity on the other. Note that all three of the reported studies employed fMRI to compare brain activations either across subjects or conditions; thus, they exemplify top-down studies resembling those cited in the memory example.

Proceeding phonetic analysis, Davis & Gaskell (2009) hypothesise left inferior frontal gyrus (IFG)—more specifically Broca's area or BA44/45—activity may reflect initial phonological interpretation while left middle temporal gyrus (MTG) and supramarginal gyrus (SMG) may be recruited for higher-level lexical processes (accessing stored phonological forms and the mental lexicon).

Largely consistent with the above suggestions, a recent model of speech perception (Hickok & Poeppel 2000, 2004, 2007) suggests that there are two functionally dissociable predominantly left-lateralized routes of speech perception from early auditory areas onwards. Hickok and Poeppel argue that the speech signal is initially analyzed with respect to its spectro-temporal features (viz. the purely auditory characteristics of speech sounds) in STG, then analyzed phonologically in posterior STS before it is propagated, in a task-dependent manner, along either a ventral "what" stream for lexico-semantic processing that connects *via* the tempo-parietal-occipital junction to distributed conceptual systems, or a dorsal "how" stream *via* inferior parietal and frontal areas mapping sounds to sensorimotor representations (possibly based on their corresponding motor commands (Liberman & Wahlen 2000)); IFG is ascribed a special role as interaction site between these routes (see figure 2.3). Note that within these processing streams there is not necessarily a unidirectional flow of information; rather, feedback connections from later to ear-

---

22 Addmittedly, it is a simplistic assumption that language comprehension can be clearly sequenced into phonological analysis of the incoming stimulus, assignment of meaning to linguistic units, and syntactic parsing of a sequence of linguistic units; but for the time being this shall suffice.

lier processing stages are abundant. Nevertheless, the majority of information is thought to propagate from early sensory areas (STS, STG) to higher cortical areas (viz. areas that are more specific to linguistic features and less dependent on the characteristics of the stimulus). The model I just introduced glosses over a number of puzzles that remain still unsolved (I will point to one such issue below) and abstracts away from small variations in the empirical results it relies on. As a consequence, it may strike some researchers as too simplistic. Yet, it offers a valuable synthesis not only of behavioral data, lesion (i.e. bottom-up) and imaging (i.e. top-down) studies but also includes evidence from computational modeling framing all of this into a bigger picture.

In cognitive neuroscience, computational models often get employed to verify hypotheses concerning the dynamics of neural interactions or the flow of activations through the neural architecture that cannot be easily assessed in living organisms. Although this is not the place to delve into the details of computational modeling approaches, note that the link between Hebb's theory of learning and the LTP mechanism discussed above (see section 2.3.1.2) is illustrative: Hebb had hypothesized that if neurons become simultaneously activated the connections between them get strengthened; although it took almost two decades to discover the biological mechanism achieving this, Hebb's theory was evidently right. More recently, artificial neural networks (ANNs) have, e.g., been employed to investigate how different clinical conditions come about as a result of degeneration in the same cortical structures. For instance, Lambon Ralph et al. (2007) successfully mimicked two types of semantic memory impairment that are, though behaviorally quite different, both associated with medial temporal lobe damage.[23] In a trained ANN, they randomly changed the weights on certain connections (viz. they changed the strength of the connections between artificial neurons) or removed the weights altogether, respectively. Subsequently, they observed different selective impairments in each of these conditions although the same initial ANN had been modified. Strikingly, the patterns of impairments found in the different conditions each matched with the observations from clinical patients suffering one of the two semantic memory impairments. This speaks to the conclusion that different kinds of damage to the same brain area can selectively affect different aspects of cognitive functioning. The modeling approach can help to clarify what exactly this connection is and thus add substantially to the localization evidence available from neuroimaging.[24]

---

[23] Patients with *semantic dementia* (SD) typically have difficulties naming subordinate categories more globally while patients with *herpes simplex virus encephalitis* (HSVE) typically suffer from category-specific impairments. For details see (Kästner 2010a).

[24] I will return to this point when discussing manipulations of *organization* in chapter 8.4 and composition studies in chapter 14.3.

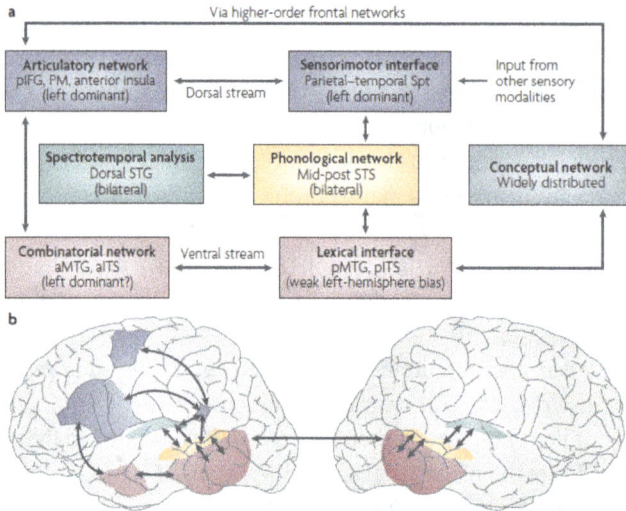

Fig. 2.3: Hickok and Poeppel's dual stream model of the functional anatomy of language. Green: spectro-temporal analysis, yellow: phonological analysis, pink: lexico-semantic "what" route, blue: sensorimotor "how" route. Reprinted by permission from Macmillan Publishers Ltd: Nature Reviews Neuroscience (Hickok & Poeppel 2007), copyright 2007.

Returning to the case of language processing, let me add that—compatible with the view advocated by Hickok and Poeppel—Ward (2006) includes the temporal pole (associated with semantic knowledge; e.g. Patterson et al. 2007) and Broca's area into the lexico-semantic "what" route while Wernicke's area (posterior BA22), the angular gyrus and, again, Broca's area form parts of the "how" stream. The hypothesized interaction-/double-role for Broca's area may be due to its involvement in working memory (WM) and monitoring processes (Friederici 2002). Whereas the lexico-semantic route is recruited when the mental lexicon is accessed, the "how" stream is recruited for access to stores of sub-lexical segments (phonemes).

At least for SpL processing, the overall picture seems reasonably clear. The existence of postulated anatomical connections has been verified using Diffusion Tensor Imaging (DTI) techniques (Saur et al. 2008, e.g.)—an MRI technique that makes visible diffusion processes of molecules (mainly water) non-invasively where diffusion is taken to reveal architectonical details. Yet, it remains unclear whether the cortical networks supporting SpL processing are indeed specialized to analyzing linguistic features or if they rather are specialized for patterning auditory input.

Given that SL shares core linguistic features of SpL, SL processing should principally engage the same cortical networks as SpL processing if the activations associated with the latter actually reflect linguistic processing. This takes us to recent research into the neural basis of SL processing.

2.3.2.2 Cortical Processing of Sign Language

In SL, information is conveyed via bodily movements—in particular movements of face, hands and arms. SL is the preferred method of communication among Deaf people.[25] Though they appear, at a first glance, drastically different from SpLs, signed languages are genuine languages as rich as any spoken language; they can be used to convey the same information as can SpLs.

Although iconic gestures are "how signs are usually born" (Teervort 1973), their iconicity tends to gradually reduce over time as signs become simplified, more regular and symmetrical (Frishberg 1975; Meier & Sandler 2004), and thus increasingly arbitrary.[26] Not only are the associations between signs and their meanings arbitrary, SLs also are productive, discrete, display compositionality, and can, just as SpLs, be used in displaced speech, i.e. to refer to objects, places, or episodes not currently present (cf. Sutton-Spence & Woll 1998). And just as with SpLs, there is a wide variety of SLs across the world, most of which are mutually unintelligible(e.g. Sutton-Spence & Woll 1998; MacSweeney et al. 2008).[27] Indeed, linguistic universals found in SpL have been identified in SL (e.g. Sandler & Lillo-Martin 2006) allowing for SLs to be subjected to the same linguistic analyses as SpLs. While it is quite obvious how this is to work for syntax and semantics, the case is slightly different for phonology.

Sandler & Lillo-Martin (2006) define phonology as "the level of linguistic analysis that organizes the medium through which language is transmitted" (p.114). In SpLs, this refers to the patterning of sounds expressed using a single articulator (sound); in SLs, it refers to the sensorimotor characteristics of signs building on multiple articulators, mainly manual ones (i.e. motion and shape of the signing hands). In both cases, there is a finite set of discrete, individually meaningless elements of features that recombine obeying certain constraints to create a lexicon of meaningful units. For SpLs, these primitive elements are *phonemes*; in SL, phonological classes for manual articulators are handshape, location, and movement (e.g. Emmorey 2002).[28]

---

25 Using "Deaf" I refer to the members of a community in terms of their language and social identification, rather than in terms of their audiological status. For the latter, I shall use "deaf".
26 Even if signs remain iconic, this makes them only analogue to SpL onomatopoeiae, viz. words imitating the sound of the object or event they represents (e.g. "tick", "splash", "glow" or "crack".
27 This is another reason why SLs cannot simply be gesture systems.
28 These categories go back to William Stokoe's (1960) seminal work.

Recognizing SLs as genuine languages allows to analyze them in a manner analogous to SpL, viz. in terms of their phonology, syntax, semantics and pragmatics. This is not to say there are no effects of modality, however. It is quite easy to see how the nature of the stimulus—an auditory compared to a visuo-spatial one—will in some way or other affect how language is processed. For instance, SLs make extensive use of physical space to express relations (SpL does not) and often use phonological elements simultaneously rather than sequentially (as does SpL). Yet, as Sandler & Lillo-Martin (2006) rightly note,

> if physical modality is responsible for certain aspects of sign language, [...] then the converse property found in spoken language [...] must also be modality induced.

Comparing and contrasting SL and SpL will thus be illuminating with respect to how languages depend on the medium through which they are conveyed.

As various experiments have shown, SL processing is generally supported by the same classical left-lateralised language areas identified for SpL processing (Capek et al. 2008; MacSweeney et al. 2004, 2008; Rönnberg et al. 2000), i.e. left IFG and left superior temporal areas (e.g. McGuire et al. 1997; Petitto et al. 2000).[29] Though most of the evidence at hand has been obtained from top-down studies (mostly functional imaging), there is bottom-up evidence in line with this: Corina et al. (1999) found stimulation to Broca's area to interfere with the motor execution of signs and stimulation to left SMG to elicit semantic and phonological errors in SL production.

Along with the perisylvian network, SL appears to recruit additional right hemispheric structures (MacSweeney et al. 2008; although see Emmorey et al. 2003; Neville et al. 1998). MacSweeney et al. (2002) found additional activation in parietal areas bilaterally and Emmorey et al. (2007) report greater activity for SL than SpL production in superior and inferior parietal lobules.

Whether these observations reflect fundamental differences between SL and SpL processing at the neural level or rather result from the visuo-spatial processing inherent in SL (Campbell et al. 2007), remains to be investigated. However, given the difference in modality it seems plausible that—at least during initial processing which is more directly dependent upon physical stimulus characteristics—more visual than auditory areas are recruited for SL than SpL.[30]

---

29 Strikingly, left-lateralisation of SL processing seems to be independent of signers' handedness (Corina et al. 2003). For a comprehensive review on SL research see MacSweeney et al. (2008).
30 Note, however, that suggesting SL may recruit areas specialised for visuo-spatial processing in addition to the classical language network, is not to say SL processing should simply be understood as 'language + visuo-spatial'.

Up to this point I have demonstrated two important things: first, SLs are genuine languages and research into the neural basis of SL processing may guide us towards a fuller understanding of the relations between certain linguistic features and the activations of specific cortical structures; second, both SL and SpL research have substantially benefitted from the integration of evidence obtained through interlevel studies—most prominently lesion and functional imaging studies. Therefore, the lesson we can draw from the above discourse adds to the conclusions drawn from the memory examples. Although (spatial) memory is the dominant example in current debates about scientific explanation, it is not cherry-picked as far as the role of interlevel studies in cognitive neuroscience goes. We find these very rationales applied also in research on cognitive phenomena that can neither be readily tracked down to specific cellular processes such as LTP nor be studied in model organisms. Admittedly, though, there is only a reduced range of bottom-up and top-down studies available compared to the memory case—but this is for practical and/or ethical reasons and it does not compromise my point.

Before closing, let me briefly sketch an ongoing cross-linguistic functional imaging study that aims to disentangle cortical processing of phonology and semantics. Looking at the concrete experimental design of one specific study will illustrate the reasoning behind interlevel studies in more detail. It is particularly interesting as it combines within and between subject comparisons in an inherently interlevel approach.

2.3.2.3 Sight & Signs

The project "Sights & Signs" is the first cross-linguistic SL study using fMRI. It employs a mixed within and between subject design to address a number of questions regarding the cortical basis of SL processing.[31] Among other things, it assesses whether cortical processing patterns identified in single SLs can indeed be generalized across SLs by comparing data from different groups of native signers. It further compares the identified networks with those known to be relevant for SpL processing. Those activations that occur across languages in the same modality as well as across modalities should, so the idea, reflect the linguistic processing inherent in both SL and SpL processing while differences may be an effect of modality. Though

---

[31] The "Sights & Signs" project is run jointly by Velia Cardin and Bencie Woll (Deafness Cognition And Language Research Centre (DCAL), University College London, England), Mary Rudner, Jerker Rönnberg, and Mikael Heimann (Linköping University, Sweden), Cheryl Clapek (University of Manchester, England), and Eleni Orfanidou (University of Crete, Greece). I shall here only report about a subset of the questions that the study addresses. For a detailed documentation of the "Sights & Signs" study see Kästner (2010b) as well as Cardin et al. (2016).

SL-SpL comparisons have been reported before, no cross-linguistic SL imaging data is available from the literature to date; this however, is crucial to have a valid basis for comparing SL to SpL.

Secondly, and more innovatively, "Sights & Signs" aims to disentangles phonological and semantic processing in SL by subtraction of conditions. Semantic processing can be isolated as the difference between activations elicited by native versus foreign signs; for both native and foreign signs are phonologically well-formed, viz. they carry phonological information, while only native signs will also convey semantic information. The phonology conveyed by handshape, location, and movement is accessible to signers only. That is, a non-signer may not realize the difference between a non-sign and a real sign while a signer will do so independently of whether or not she can assign a meaning to the sign in question. This is because the specific combinations of handshape, movement and location permissible in SL are largely consistent across SLs—they *make sense* to signers. This is different for non-signs that are designed in such a way that—although using similar handshapes, movements, and locations as real signs—they are phonologically ill-formed violating phonotactic constraints. To signers, these non-signs simply *look weird*; they cannot be subjected to phonological analysis (see figure 2.4 for two examples). The comparison between foreign sign and non-sign processing in signers will thus isolate phonological SL processing. Note that these comparisons are unavailable in SpL research: since all SpLs are based on some sort of sound, non-words for SpL can still be subjected to phonological analysis (i.e. parsed into letters or sounds) while non-signs in SL cannot. Likewise, since all signs are phonologically meaningless to non-signers, we can compare foreign sign processing across groups (signers vs. non-signers) to isolate phonological SL processing.

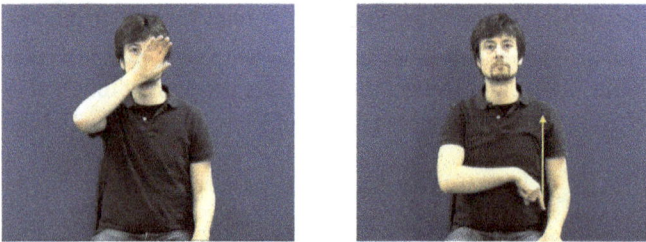

Fig. 2.4: Left: non-sign occluding part of the signer's face. Right: non-sign with phonologically implausible upward movement.

Addressing the aforementioned questions requires looking at a minimum of two mutually unintelligible SLs that have unique as well as shared signs. For "Sights & Signs", British Sign Language (BSL) and Swedish Sign Language (SSL) have been selected as these languages bear sufficiently little resemblance despite a number of cognates. To assess linguistic processing and distinguish it from visuo-spatial processing inherent in SL, not only signing groups but also a control group of non-signers is needed where both groups will be presented with identical stimulus material and tasks.[32] Subtraction of non-signers' cortical activations from signers' cortical activations will reveal the linguistic components of cortical SL processing. Similarly, a baseline condition can be used to isolate cortical activations that are directly stimulus-induced from other (default or background) activations. Figure 2.5 schematically depicts the experimental design used in "Sights & Signs".[33] Note that throughout the experiment, all participants are asked to perform two different tasks—a handshape and a location matching task—on a subset of the stimulus material. In both tasks, participants were shown a cue of a handshape or location prior to each block of stimulus items. They were ask to press a button as soon as they spotted the cued handshape or location, respectively. For current purposes both these tasks can just be seen as a means to ensure participants remain awake and alert and actually process the information displayed to them.

Given this setup, and idealizing away from methodological complications, we can interpret "Sights & Signs" as a top-down study in the following sense: the change between conditions, viz. from one type of stimulus to another, serves as a manipulation of participants' language processing. This is because which type of linguistic information is available to our participants immediately determines whether or not participants engage in phonological and/or semantic processing. At least if we assume that they are awake and alert, engage in the task, try to process the linguistic information available but do not try do to this where there is no linguistic information available for them. Admittedly, these assumptions weigh quite heavily but even if they do not hold, this does not affect the experimental logic *in principle* but just is a complication for this concrete case. Besides, there is not much we can do without these assumptions within the current setup.

Now, cortical activations are measured across all conditions in all subjects throughout the experiment. Once we compare neural activations during non-sign and native sign presentation, we observe significantly more activation in (primarily

---

32 For "Sights & Signs", we used handshape and a location matching tasks.

33 I shall skip the minute details of participant and stimulus selection as well as the technical aspects of scanning and data analysis; but be assured that everything is done to in accordance with the highest standards of experimental research. For a detailed documentation see Kästner (2010b).

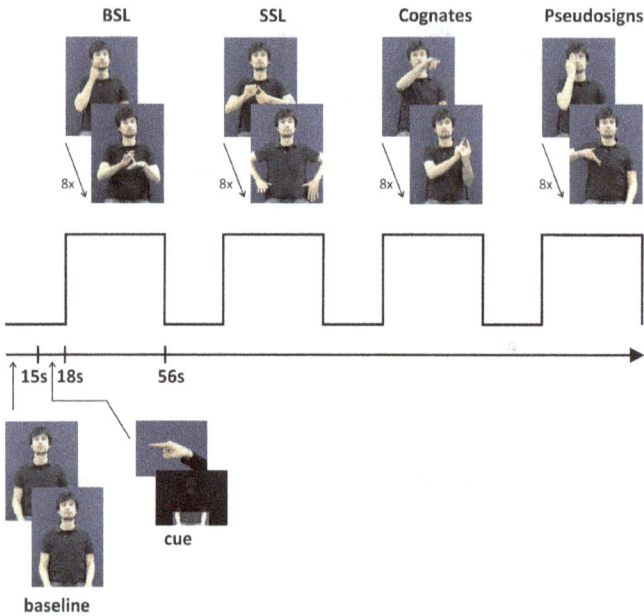

Fig. 2.5: Experimental design for "Sights & Signs".

left) superior temporal gyrus (STG), superior temporal sulcus (STS), inferior fronto-parietal areas, the temporal-parietal junction (TPJ), and anterior temporal areas for native signs. Put slightly differently, the activations in these areas increased as an effect of changing from non-sign to native sign presentation. Analogously, when we contrast foreign to native sign presentation, we observe a significant increase of activations in (again, primarily left) inferior fronto-parietal areas, TPJ, and anterior temporal areas for native signs. Both comparisons can thus be read as classical top-down studies: the manipulation of linguistic (phonological and/or semantic) processing elicits changes in cortical activation patterns associated with it.

Contrasting the results from these two studies, we can conclude that phonological processing must be correlated with (primarily left) STS and STG activations. This is because these regions only showed an increase in activation when changing from non-signs to native signs (i.e. adding semantic and phonological information) but not when changing from foreign signs to native signs (i.e. adding both semantic information only).

We should expect that this result can—in principle—be confirmed in bottom-up study, e.g. using transcranial magnetic stimulation (TMS). When TMS is applied to neural tissue (usually through the skull), it induces a current that temporally disrupts all signal processing in the area it is applied to. This allows researchers to temporarily "knock out" specific brain regions creating a "virtual patient". To complement the results from the sign language study described above, TMS could be applied to left STS and STG. As a result, we would expect participants to fail in phonological tasks directly following TMS application. That is, manipulating neural processing in left superior temporal areas interferes with phonological processing.

2.4 A Matter of Difference Making

Taken together, both the case of memory and the case of language processing show how a systematic combination of different interlevel experiments tapping into the same cognitive phenomenon has advanced our understanding of how that cognitive phenomenon is grounded in the neural workings of biological brains. Importantly, the experimental logic I have been outlining is largely independent of the cognitive phenomenon under investigation. It is so abundant indeed that I could well choose different examples.

We have looked at the role of LTP and hippocampus in spatial and recognition memory, as well as functionally specific cortical oscillations. In a second step, we have left the realm of memory research and moved to language processing—a phenomenon for which, unlike memory, very little is known about the molecular mechanisms that underlie it and research focuses much on localization issues instead. Even in the very specialized field of sign language research, we identified the very same patterns of experimental practices utilizing difference making in the context of dissociations and interlevel studies.

Most of the studies I discussed crucially build on *difference making*: scientists fiddle with one thing to observe whether or not that makes a difference to another. This is the intuition behind both reasoning about dissociations and interlevel experiments—or at least this is how the stories are usually told. In fact, though, this way of telling the story is highly idealized: there is much more to scientific experimentation and empirical manipulation than manipulating one thing and seeing to which other things it makes a difference. I will return to this point in chapters 12 and 14. For now, the lesson to learn is that discovery of dependencies and independencies between cognitive and neural phenomena is an important cornerstone of cognitive neuroscience. Knowing what does and what does not affect (or is and is not affected by) a phenomenon helps us understand how that phenomenon is produced (or what its role with respect to another phenomenon is). And this general principle probably holds true for many other (special) sciences as well.

2.5 Upshot

Traditionally, investigating mental, or cognitive, phenomena has been the domain of psychology—an experimental science that itself only emerged in the late 19th century. However, scientists approach to studying how cognitive phenomena come about has been transformed over the past century. Most peculiarly, the focus has shifted from merely studying cognitive phenomena to investigating cognitive processes alongside their neural substrates. Facilitated by the availability of powerful functional imaging techniques, a new discipline has emerged:*cognitive neuroscience*. As a result of this transformation of empirical research into cognitive phenomena, contemporary explanations often cite neural structures and processes to be responsible for their production. And indeed, cognitive neuroscientists have successfully advanced our understanding of how some cognitive phenomena are produced by the brain. Therefore, if we want to understand how scientists explain cognitive phenomena, this is the place to look.

Cognitive neuroscience as it is practiced today is a relatively young, highly specialized, yet interdisciplinary, research area that employs a large pool of different methodologies. It studies a wide range of phenomena and uses a plurality of descriptions, especially psychological and neural ones. The prominence of interlevel investigations within this field is what makes cognitive neuroscience the perhaps most crucial discipline within contemporary cognitive science. For when it comes to the question of what the neural bases, substrates, or realizers of (human) cognitive functions are and how cognitive processes or phenomena come about, we eventually want to account for cognitive processes in terms of neural ones.

This places cognitive neuroscience in a unique position. First, it incorporates aspects of many different special sciences due to the expressed methodological variety in the field. Second, it presents some peculiar challenges about relating different levels or domains (neural, cognitive) of investigation in the construction of scientific explanations. Thus, if we can develop an account of scientific explanation that meets the demands of cognitive neuroscience, this account is likely to generalize to other special sciences or, at the very least, function as a paradigmatic illustration for what challenges contemporary philosophy of science faces and how they might be met.

Excursus: Dissociation Logic

In psychology and cognitive neuroscience, a dissociation is defined as a patient's selective functional impairment on a particular cognitive task. Most generally, there are two kinds of dissociations permitting different inferences. In a *single dissociation*, only performance on one particular task (say, task A) is impaired while performance on another (say, task B) is relatively spared. For illustration, consider two observed patients: P1 suffers from a lesion to structure X and her performance on both tasks A and B is impaired (no dissociation); P2 suffers from a lesion to structure Y but her performance is impaired only on task A, not on task B. In this case (assuming functional and architectural equivalence of P1 and P2), we infer that task A requires structure Y to be intact while task B does not.

Where two complementary single dissociations are found in two different patients, neuropsychologists speak of a *double dissociation*: if P1 in the above case was selectively impaired on task B but not A while the opposite was true for P2, and P1 performs task A better than P2 while the reverse holds for task B, neuropsychologists infer that performing task A requires structure Y while B requires X. This would falsify any scientific theory according to which performing tasks A and B relies on the same cortical (sub-)system (e.g. Ward 2006). Note, however, that reasoning about dissociations as such does not necessarily require the association of cognitive functions with particular brain areas; dissociations can be based merely on behavioral evidence that gets used to make inferences about the relations of psychological capacities and develop psychological models.

Yet, the rise of cognitive neuroscience in the late 20th century and the development of powerful neuroimaging methodologies together with the conviction that cognitive processing is biologically grounded in the brain, has led to a tight connection between reasoning about cognitive functions and the neural structures that are believed to implement them. This is especially evident in cognitive neuroscience—a discipline that already by name combines studying the cognitive with studying the neural domain.

3 The Life of Mechanisms

Summary
Philosophy of science has recently experienced a mechanistic turn: while the received view used to be that scientific explanations conform to the deductive-nomological (D-N) model, mechanistic explanations are now up and coming. According to the D-N model, to explain a phenomenon is to subsume it under a law. According the mechanistic view, to explain a phenomenon is to identify the mechanism, and the internal workings thereof, that brings the phenomenon about. The current chapter introduces mechanistic explanations and puts them into a historical as well as an empirical research context.

3.1 Ancestors: Some Historical Context

Mechanical philosophy has been prominent in the 17th century. Robert Boyle (1627-1691), for instance, proposed a *corpuscularian theory* according to which all properties of macroscopic objects, even large-scale phenomena, are determined by the properties and interactions of their composing corpuscles—tiny bodies with a fixed size and shape and varying degrees of motion. Although this theory bears remarkable resemblance to contemporary mechanistic views, mechanisms have not received much attention in the philosophy of science between then and the millenium.

Rather, the received view of explanation in philosophy of science up until the 1980s has been the deductive-nomological (D-N) account. According to the D-N model, explanation works essentially by deductive reasoning along the following schema:

(1) $\forall x : (Px \supset Qx)$

(2) Pa

---

(3) Qa

In the above example, Qa is the explanandum phenomenon. In order to explain Qa, we can appeal to the general law $\forall x : (Px \supset Qx)$ and the observation of Pa. Together they form the *explanans*. According to Carl Hempel and Paul Oppenheim's seminal formulation of the D-N account, the explanandum is successfully explained by the explanans only if a number of conditions are satisfied. Most importantly, all

DOI 10.1515/9783110530940-003

sentences in the explanans have to be true and the explanandum has to be a logical consequence of the explanans (Hempel & Oppenheim 1948).

The turn back towards a more mechanistically inclined approach to scientific explanations came in the 1980s; it was probably initiated by Wesley Salmon's work. After much emphasis had been placed on laws, logic, and deduction, and philosophy of science had primarily been tailored to physics, Salmon realized that scientists did not typically deduce phenomena from laws. The time had come to "put the 'cause' back into 'because'" (Salmon 1984, p. 96). After all, explanations should tell us something about the causal workings underlying the phenomenon to be explained.

According to Salmon's causal-mechanical (C-M) model (Salmon 1984, 1989), explanations are not arguments but descriptions of causal processes in the world. Causal processes are, on Salmon's view, essentially physical processes transferring energy and momentum from cause to effect or transmitting a *mark*, viz. some kind of local modification of a structure or process. Any process that would transmit a mark, if it was appropriately marked, is to be considered causal. As his prime example Salmon uses radioactive tracers that can be used to track certain physiological processes and thus help us observe how a certain substance, e.g. a specific kind of transmitter, moves through a body. On the C-M view, scientific explanations trace both the continuous causal processes leading up to the event to be explained and the processes and interactions that make up the event itself; Salmon calls these the etiological and constitutive aspects of explanation, respectively. An explanation of a given phenomenon thus shows us how it "fit[s] into the causal nexus" (Salmon 1984, p. 9) of the world.

There are problems with the C-M account, however. First, it lacks a workable account of *relevance*: even if Salmon is right that we can identify causal processes by some kind of physical contact or mark transmission, this does not ensure those causal processes are actually relevant to the phenomenon to be explained.[1] Second, Salmon's notion of causation seems somewhat too strict: there are, after all, cases of *action at a distance* that we do want to render causal. The gravitational interaction between the earth and the sun, for instance. On Salmon's view, we would have to render such interactions non-causal and non-explanatory because no physical mark is being transmitted. Yet, there is clearly something causal and explanatory about it.

---

[1] Consider the case of John taking his wife's birth control pills. John's body will certainly ingest and process the substances prohibiting pregnancy in fertile women. Similar (albeit different) continuous spatio-temporal processes will be at work when John's wife takes birth control pills. Nevertheless, the pills are relevant only to the non-pregnancy of John's wife, not to John's own non-pregnancy.

Following Salmon's example but recognizing the problems associated with the C-M account, philosophers of science began to discuss about causation more vigorously. Motivated by the goal to provide a convincing theory of causation and scientific (causal) explanation—ideally one that is in line with empirical research practices—two new theories developed: *mechanistic explanations* and the interventionist account of causation, *interventionism* for short. The current chapter focuses on mechanisms, the following chapter introduces the interventionist view.

3.2 The Birth of Mechanistic Explanations

It is perhaps Stuart Glennan's 1996 paper that marks the beginning of contemporary discussions of mechanisms in the context of scientific explanation. Glennan's motivation for providing a theory of mechanisms is to base a theory of causation upon it. This, in turn, he is motivated to do by Hume's argument to the effect that whilst we can observe some event regularly following another, we may still never see the *secret connexion* between the two (cf. Hume 1902 [1777]). Hume's secret connexion, Glennan argues, is not actually secret; it can be laid bare once we consider the *mechanism* that connects the two events. But what *is* a mechanism? Glennan provides the following definition:

> A mechanism underlying a behavior is a complex system which produces that behavior by the interaction of a number of parts according to direct causal laws. (Glennan 1996, p.49)

A mechanism as Glennan describes it is as a sequence of causal processes, governed by direct causal laws, leading from one event to another. Mechanisms can produce certain outputs in response to certain inputs (like a vending machine) or produce periodic behavior (like circadian rhythms).

Glennan considers the example of the turning of a car key being followed by the starting of the engine: we can reveal the mechanism connecting key-turning and engine-starting by looking under the car's hood.[2] This mechanism, Glennan holds, is what *explains* that the key-turning is followed by the engine-starting. According to Glennan, the principle idea of a mechanism has a very wide range of application; it applies "to all sciences except fundamental physics" (Glennan 1996, p.56). Yet, he admits, there are regularities that cannot be accounted for in terms of discernible mechanisms such as the gravitational attraction of bodies.

---

2 Glennan explicitly points out that it need not be oneself personally who can figure out what is happening under the hood. It is sufficient if anyone (e.g. a car expert) can do so.

Though inspired by 17th century mechanical philosophy, Glennan significantly departs from it. For one thing, he allows mechanisms to have simple and complex parts, not just fundamental particles. For another, he does not restrict interactions between mechanism parts to physical ones. Instead, what matters for an interaction to be explanatory is that an expert can, in principle, access what is happening when the interaction in question is taking place. Precisely what, though, does Glennan have in mind when he speaks of interactions?

Glennan's notion of interaction clearly is a causal one. Interactions between a mechanism's parts are to be governed by *direct causal laws*. Laws, for Glennan, are universal counterfactual-supporting causal generalizations. They are *direct* insofar as they involve no intervening parts but only the parts immediately related to one another in the law such that "the behavior of the aggregate stems from a series of local interactions between parts" (Glennan 1996, p. 56).[3] This highlights, once again, Glennan's conception of mechanisms as causal chains. Indeed, he equates the two when he states that "a relation between two events [...] is causal when and only when these events are connected in the appropriate way by a mechanism" (Glennan 1996, p. 56). This turns the problem of distinguishing between causal connections and mere correlation into a scientific one:

> If one can formulate and confirm a theory that postulates a mechanism connecting two events, then one has produced evidence that these events are causally connected. The necessity that distinguishes [causal] connections from accidental conjunctions [i.e. correlations] is to be understood as deriving from [the] underlying mechanism, where the existence and nature of such a mechanism is open to investigation. (Glennan 1996, p. 64)[4]

On Glennan's view, there are two kinds of direct causal laws: *mechanically explicable* and *fundamental* ones. While fundamental laws "represent facts about which no further explanation is possible" (Glennan 1996, p. 61), mechanically explicable laws are "realized by some lower-level mechanism" such that "every instance of a non-fundamental law is explained by the behavior of *some* mechanism" (Glennan 1996, p. 62, emphasis in original). Mechanisms thus form some kind of hierarchy that bottoms-out where fundamental physics is reached.

Two aspects of Glennan's view are particularly interesting in the current context. First, he does not hold onto the traditional view of causation being something that operates solely on the level of fundamental physical particles. Rather, he pos-

---

3 In this context "local" is not to be understood spatially. Instead, it means there are no intervening parts or "immediate actor on the next" (Tabery 2004, p. 3). Thus, Glennan can also handle cases of causation at a distance.

4 An exception, according to Glennan, are causal relations in fundamental physics that cannot be further explicated.

tulates causal relations at every level of mechanistic of analysis. Second, Glennan defines a mechanism with respect to an underlying behavior, not a mechanism *simpliciter*. This is, he contends, because a mechanism is always to be identified relative to what it does, it is always a mechanism *for* something:

> One cannot [...] identify a mechanism without saying what it is that the mechanism does. The boundaries of the system, the division of it into parts, and the relevant modes of interaction between these parts depend upon what [is] the behavior we seek to explain. [...] If one isolates a complex system by some kind of physical description, one can identify indefinitely many behaviors of that system [...]. A complex system has many mechanisms underlying its different behaviors. (Glennan 1996, p. 52)

That is to say exactly how we divide up the system to explain a given behavior very much depends on what precisely that behavior is (see also Wimsatt 1972). It may further be influenced by the particular view we take at a system, what tools and methodologies we are using to investigate it, or what our disciplinary background is. Just as the mechanism itself, its parts can be subjected to mechanistic analysis. Likewise, a mechanism can be placed into a larger context forming itself part of a more complex system. This illustrates how we can arrive at nested hierarchies of mechanisms when analyzing a complex system.

In summary, Glennan's theory already explicates some key ideas of modern mechanistic views (as we shall see below). It revives aspects of mechanical philosophy and also shows clear footprints from the positivist tradition. Despite the focus on laws Glennan goes beyond D-N explanation and follows Salmon in bringing causality back into the picture. His approach is quite different from Salmon's C-M model, though: rather than being restricted to physical interactions, Glennan adopts a more liberal view according to which all kinds of mechanical interactions are causal processes, even those characterized at higher mechanistic levels or those involving action at a distance.

While both Glennan and Salmon tightly interlinked their accounts of explanation with accounts of causality, modern discussions about mechanistic explanations have become somewhat separated from accounts of causality. Yet, as we will come to see in parts II and III, discourses on mechanisms and causation merge back together. For now, let us turn to the contemporary mechanistic view.

3.3 Upbringing Around the Millennium

At least since the turn of the millennium, mechanisms have received increasing attention in philosophy of science. Peter Machamer, Lindley Darden and Carl Craver's (2000) seminal paper "Thinking about Mechanisms" probably is the most-cited paper on mechanistic explanations. However, they were not the only ones to advance mechanistic accounts around the turn of the century. In 2002, James Woodward offered a counterfactual account of mechanisms and Glennan published an account incorporating some of Woodward's ideas. In 2004, Machamer advanced a revised characterization of mechanisms concerned more closely with the metaphysics and epistemology of mechanisms. William Bechtel and Adele Abrahamsen (2005) picked up mechanistic explanations in the context of scientific reasoning. The probably most detailed account of mechanistic explanation today was presented by Craver (2007b) in his book "Explaining the Brain". While there are many parallel developments linking the different mechanistic accounts, my discussion here must be limited. I will begin by introducing Machamer, Darden and Craver's initial account of mechanisms, then outline Woodward's and Glennan's views. Finally, I will turn to Craver's own account. Later chapters will mostly refer to this as the received mechanistic view. Yet, it will also be important for the discussions in this book to see how Craver's view compares and contrasts with other mechanistic accounts.[5] Some of the open questions and issues that will be central to the discussions in later chapters will be sketched in due course.

3.3.1 Machamer, Darden & Craver

Machamer, Darden and Craver (2000) (MDC hereafter) focus mostly on explanations in neurobiology and molecular biology. According to their view, mechanisms are *composed of* entities (having certain properties) and activities:

> Mechanisms are entities and activities organized such that they are productive of regular changes from start or set-up to finish or termination conditions. (MDC, p. 3)

Within a mechanism, each entity is playing a specific role. That is to say entities have specific functions that "should be understood in terms of the activities by virtue of which entities contribute to the workings of a mechanism." (MDC, p. 6) These "activities are ways of acting, processes, or behaviors; they are active rather than passive; dynamic rather than static" (MDC, p. 29). Activities on MDC's account are appropriately compared to Salmon's interactions. There are several different classes of activities: geometrico-mechanical, electro-chemical, energetic, and electro-magnetic ones. However, they are not as restricted: "Mere talk of transmission of a mark or exchange of a conserved quantity does not exhaust what

---

5 For recent elaborations on Craver's view also see Craver & Darden (2013); Craver & Tabery (2015); Craver (2015).

[...] scientists know about productive activities and about how activities effect regular changes in mechanisms." (MDC, p. 7)

On MDC's view, entities and activities are interdependent: certain entities can engage in certain activities and certain activities can be performed by certain entities; activities are "the producers of change" (MDC, p. 3), they are "types of causes" (MDC, p. 6). When an entity engages in a productive activity it acts as a cause; but it is not the entity *simpliciter* that is a cause.[6] Activities constitute the continuous transformations that lead from startup to termination conditions of a mechanism.

Like Glennan, MDC hold that mechanisms are always mechanisms *for* something (a behavior, a phenomenon) we want to explain. They are "identified and individuated by the activities and entities that constitute them, by their start and finish conditions, and by their functional roles." (MDC, p. 6) Entities in mechanisms are, in turn, identified and individuated by their properties and spatiotemporal locations. Activities may also be identified and individuated by their spatiotemporal locations as well as features like rate, duration, mode of operation, etc. (cf. MDC, p. 5). Explaining a phenomenon for MDC is to give a description of how it was *produced* by the particular organization of entities engaging in a continuous series of activities composing a mechanism. To produce a specific phenomenon, entities need to be appropriately located, structured, and orientated while the activities in which they engage must have a temporal order, rate, and duration. Given stable circumstances, a mechanism is expected to always (or for the most part at least) work in the same way; in this sense, mechanisms on MDC's view are *regular*.[7]

According to MDC, mechanistic descriptions can be provided at multiple levels and linked up with one another:

> Mechanisms occur in nested hierarchies and the descriptions of mechanisms in neurobiology and molecular biology are frequently multi-level. The levels in these hierarchies should be thought of as part-whole hierarchies with the additional restriction that lower level entities, properties, and activities are components in mechanisms that produce higher level phenomena [...]. (MDC, p. 13)

However, it will not always be useful—let alone possible—to unfold the complete hierarchy of mechanisms involved in bringing about a phenomenon. Therefore, we may find varying degrees of abstraction in mechanistic explanations, depending on how much detail is included; we may at times find abstract terms used to capture

---

6 MDC provide no clear definition of what it is to be a cause. They treat "cause" as a generic term that one cannot (and need not) have a theory *tout court* of (see Machamer 2004).
7 Machamer (2004) suggests to drop the qualification "regular" from MDC's definition of a mechanism.

complex activities in higher-level mechanisms that we could further unfold. Such an abstract description of a mechanism that can be filled in with known parts and activities MDC call a *mechanism schema*. Mechanism schemata are different from *mechanism sketches* that contain missing pieces and black boxes such that we cannot fill it all in to yield a mechanistic explanation.

The overall tune of MDC's account is in line with that presented by Glennan's. Both accounts share Salmon's view that explanations refer to causal processes operating over time.[8] Mechanisms, according to both MDC and Glennan, are *decomposable* systems defined relative to a given phenomenon or behavior we want to explain (the explanandum). Their parts do something such that a continuous chain of causal processes leads up from one event to another.[9] To provide an explanation thus means to cite the mechanism's parts and how they causally operate.

What counts as an appropriate explanation is, according to both Glennan and MDC, interest-relative. Mechanistic descriptions build a nested hierarchy and explantations can be given at many different "levels" (though neither MDC nor Glennan explicate this term in detail). At some point, this hierarchy will bottom-out. While Glennan takes this to happen once the laws of fundamental physics are reached, MDC state that the lowest possible level too depends on scientists' interests and available methodologies, etc.

There are several points on which MDC's and Glennan's views diverge more crucially. For instance, MDC appeal to startup and termination conditions in their definition which Glennan does not mention. Further, MDC take a conspicuously functionalist stance on the identification and individuation of mechanistic components which Glennan does not express—or at least not as vigorously. Third, while Glennan explicitly invokes an analogy with machines such as a cars, MDC essentially understand mechanisms as active things operating by themselves and quite unlike a construction that will be passively used.[10] Finally, and perhaps most imporatantly in the current context, MDC do not appeal to laws. Indeed, they openly criticize Glennan for doing so:

> We find the reliance on the concept of "law" problematic because, in our examples, there are rarely "direct causal laws" to characterize how activities operate. (MDC, p. 4)

---

8 MDC even state explicitly that it is "appropriate to compare [their] talk of activities with Salmon's talk of interactions." (p. 21).

9 The appeal to a system's decomposability follows a long tradition (cf. Wimsatt 1976; Bechtel & Richardson 1993).

10 This may just be because Glennan focuses on man-made artifacts while MDC focus on biological mechanisms.

This arguably makes MDC's view much more apt for scientific explanations outside physics. For appealing to strict exceptionless laws does not seem plausible in an exception-prone empirical context. However, as we will see below, Glennan later argues that he did not actually think of laws as strict, exceptionless rules but rather as change-relating generalizations (see section 3.3.3).

3.3.2 Woodward

Much in line with MDC, James Woodward presented a counterfactual account of what a mechanism is in 2002. He also suggests that mechanisms consist of parts, but adds they must be *modular*, viz. that mechanism parts must be independently modifiable.[11] Similar to Glennan and MDC, Woodward fashions mechanisms as an account of causation and causal explanation. What we aim for when explaining a phenomenon is a causal story rather than a descriptive one. We want it to tell us how the mechanism's parts are linked to one another and what we need to manipulate to bring about a change in the phenomenon under investigation. Like MDC, Woodward holds that mechanisms exhibit some kind of *regularity* and that they are *productive* where productive relationships are causal in nature. Also, Woodward sides with MDC in criticizing Glennan's appeal to direct causal laws; he states that the appeal to laws is "of very limited usefulness in characterizing the operation of most mechanisms." (Woodward 2002, S368) Rather than holding universally, generalizations describing mechanisms hold approximately or under certain background conditions only.

However, Woodward is not satisfied with MDC's alternative characterization of production in terms of different classes of activities (geometrico-mechanical, electro-chemical, energetic, and electro-magnetic) either. Something stronger is needed, he contends, to do the work of Glennan's laws. His solution is that a mechanism's behavior "conforms to generalizations that are invariant under interventions" (Woodward 2002, S366). What does this mean precisely? Woodward's analysis is based on the concept of *invariant generalizations*:

> if a generalization G relating X to Y is to describe a causal relationship (or in MDC's language) a "productive" relationship, it must be invariant under at least some (but not necessarily all) interventions on X in the sense that G should continue to hold (or to hold approximately) under such interventions. (Woodward 2002, S370)

---

11 As we will see in due course Woodward's interventionist account of causation (see chapter 4) heavily resembles his account of mechanisms presented here. The modularity requirement will return as (CMC) and (IF).

To grasp Woodward's concept of invariant generalizations, we need to know what he means by *intervention*. Basically, we can think about interventions as idealized experimental manipulations.

> Suppose that X and Y are variables that can take at least two values. The notion of an intervention attempts to capture, in non-anthropomorphic language that makes no reference to notions like human agency, the conditions that would need to be met in an ideal experimental manipulation of X performed for the purpose of determining whether X causes Y. The intuitive idea is that an intervention on X with respect to Y is a change in the value of X that changes Y, if at all, only via a route that goes through X and not in some other way. This requires, among other things, that the intervention not be correlated with other causes of Y except for those causes of Y (if any) that are causally between X and Y and that the intervention not affect Y independently of X. (Woodward 2002, S370)

Woodward's idea is that appropriate interventions will allow us to sort accidental from causal generalizations. Since we want explanations to tell us a causal story, this is vital for the explanatory enterprise and to constructing mechanistic explanations.[12] Based on this insight, Woodward gives the following counterfactual characterization of mechanisms:

> a necessary condition for a representation to be an acceptable model of a mechanism is that the representation (i) describe an organized or structured set of parts or components, where (ii) the behavior of each component is described by a generalization that is invariant under interventions, and where (iii) the generalizations governing each component are also independently changeable, and where (iv) the representation allows us to see how, in virtue of (i), (ii) and (iii), the overall output of the mechanism will vary under manipulation of the input to each component and changes in the components themselves. (Woodward 2002, S375)

Appealing to invariant generalizations for identifying causal relations not only seems more adequate than MDC's "production of regular changes" or Glennan's laws but also trumps over Salmon's view according to which causal production is a matter of transferring conserved quantities. To argue for this conclusion, Woodward considers an example of *causation by disconnection*: the lac operon in E. coli bacteria.

Lactose metabolism in E. coli bacteria is regulated by enzymes. Enzyme production, in turn, is regulated by gene transcription. According to François Jacob and Jacques Monod (1961), there is an operator region controlling whether or not the structural genes coding relevant enzymes can be transcribed. As long as no

---

12 The idea is based on Woodward (2000). Woodward's theories about causation and causal explanation culminated in his seminal book "Making Things Happen" (2003) where he defends an interventionist account of causation (see chapter 4).

lactose is present, a regulatory gene produces a repressor which binds to the operator thus preventing gene transcription. If lactose is present, allolactose which is formed from lactose binds to the repressor, inactivating it and thus giving free reins to gene transcription and enzyme production.

In this example, we clearly face a mechanism (on Woodward's, MDC's and Glennan's views). And there clearly seems to be a causal connection between the presence of allolactose and enzyme production which involves the removal of the repressor. If disrupting a physical connection is causally relevant, causation is *contra* Salmon not adequately characterized as a process of physical mark transmission. Also, *contra* Glennan the connection between repressor removal and enzyme production cannot be characterized in terms of strict exceptionless laws because there will be exceptions in this biological mechanism. Finally, since the non-activity of the repressor causes enzyme production the E. coli case suggests that non-activities can be causes. Since MDC suggest activities to be causes, non-activities should not be, Woodward argues. However, Machamer (2004) counters that as long as a non-activity goes along with some activity elsewhere in the mechanism in question, there is no problem in attributing productive (causal) status to this non-activity. Independently of whether or not this is convincing, Woodward's invariant generalizations do seem to have a much wider range of application than MDC's activities as they can capture any kind of changes, not only physical ones.

Before turning to the next step in the development of mechanistic explanations, let me point out an interesting feature of Woodward's account that will become important in due course (see chapter 7): according to his definition, interventions serve two related but different purposes. For one thing, Woodward appeals to invariance and interventions to pick out mechanistic parts, to establish their independent modifiability (modularity) and to uncover the causal relations between them. For another, invariance and interventions get employed to understand and explain how a mechanism produces an overall phenomenon.

3.3.3 Glennan Again

Attracted by Woodward's notion of invariant generalizations, Glennan (Glennan 2002, 2005) presents a revised account of mechanisms in which he basically replaces the "causal laws" form his 1996 definition with "invariant, change-relating generalizations".

> A mechanism for a behavior is a complex system that produces that behavior by the interaction of a number of parts where the interactions between parts can be characterized by direct, invariant, change-relating generalizations. (Glennan 2002, S344)

Glennan is quite explicit that his revision is to be seen merely as a change in wording; he never meant to postulate strict nomic connections and contends that "whether we choose to call them laws or something else, mechanists of all stripes agree that there are non-accidental, but not truly exceptionless generalizations that describe the behavior of mechanisms." (Glennan 2010b, p. 368) Interestingly though, Glennan still holds onto some kind of positivist-fashioned laws. In addition to those *mechanically explicable laws* (as he calls Woodwardian generalizations), Glennan postulates a second kind of laws: not mechanically explicable ones which represent "brute nomological facts about the universe" (Glennan 2002, S348). They are probably found at the bottom of the mechanistic hierarchy.

As in his 1996 conception of mechanisms, Glennan emphasizes that mechanisms consist of parts. Parts are spatially-localized physical entities that *directly interact* in the sense that there are no intervening parts. As mechanisms form a nested hierarchy, mechanisms can contain (lower-level) mechanisms as parts. What explains a given behavior is the chain of causal interactions within the mechanism producing this behavior where parts in this causal chain may themselves unfold into mechanisms. Despite the talk about hierarchies, even a fully unfolded mechanistic explanation will—on this conception—essentially be a chain of causal interactions between (lower-level) mechanism parts.

3.3.4 Activities vs. Interactions

As already mentioned earlier, Glennan's and MDC's conceptions of mechanisms share many important features; this has not changed with Glennan adopting Woodward's invariant generalizations. On both views, mechanistic explanations describe the mechanism's behavior in terms of the causal processes underlying the explanandum phenomenon. Yet, both views place different emphases on different aspects of scientific explanation. While Glennan places himself mostly in the context of traditional philosophical debates on causation and explanation, MDC are much more interested in actual empirical practice. This is probably why their views diverge in an important point, viz. how the behavior of mechanistic parts is conceptualized.

On Glennan's view, mechanisms operate by their parts interacting; where an interaction is a property change in one part that results from a property change in another part; this is reminiscent of Salmon's mark transmission.[13] On MDC's view, however, mechanisms are collections of entities and activities that "produce"

---

13 Later, in Glennan (2010b) this is even more explicit.

regular changes. Their ontology thus requires both entities and activities and not just entities having (and changing) properties. This might be advantageous; for it allows us to individuate and refer to activities independent of any particular entity (cf. Machamer 2004). After all, there are different kinds of activities just as there are different kinds of entities. The same activity may be performed by different entities and the same entity may perform different activities.[14]

Since Glennan and MDC seem to offer interactions and activities, respectively, as a replacement for Salmon's mark transmission, we might think that these notions are in competition. But perhaps they are not. First, activities might not get off the ground without interactions. James Tabery offers the following diagnosis:

> the minute one starts to examine in what sense these activities are to be productive in a mechanism, then some notion of Glennan's interaction as an occasion whereby a property change in one entity brings about a property change in another entity is required alongside of the concept of an activity. (Tabery 2004, p. 8)

This is to say that we cannot analyze activities without a clear understanding of interactions. I agree this might be true for some activities, e.g. activities involving multiple entities. But I doubt it will be for all activities. For activities that are just a matter of a single entity doing something, say a radioactive molecule decaying, there might not be need for interactions at all.[15] Second, MDC's activities seem to capture more then just the causal processes connecting mechanistic parts: they also somehow "make up" or "constitute" the behavior of the mechanism as a whole.[16] This is something Glennan's interactions do not do. In fact, Glennan does not have to bother with such hierarchical relations of making up. Mechanisms on MDC's view form part-whole hierarchies that are quite different from Glennan's nested hierarchies. Glennan's hierarchies will eventually unfold into a single chain of causal processes—somewhat like a nested formula unfolds into a single chain of mathematical operations once all the bracketed expressions are written out. MDC's hierarchies, on the other hand, are hierarchies of composition. Unfolding these will be more like taking apart a Matryoshka doll.

---

14 See Illari & Williamson (2012) for an argument to the effect that activities should be included in a mechanistic view.

15 While I do not aim to offer a metaphysical analysis of mechanisms, I will briefly return to the discussion of activities, interactions, and production in chapter 12.3.

16 Indeed, "productivity" on MDC's view also seems to capture some kind of interlevel "making up" relation where the individual activities of entities *make up* or *produce* the overall behavior of the mechanism. I will get back to this in chapter 13.

3.4 Growing Famous

The perhaps most elaborate account of mechanistic explanations has been proposed by Carl Craver. In his seminal book "Explaining the Brain" (2007), he focuses on explanations in neuroscience. Though the view he advocates is similar to MDC's it is not quite the same; we will see this in due course. As Craver sees it, scientists are looking for the (neural) mechanisms implementing the (cognitive) phenomena they are trying to explain. He suggests that explanations in neuroscience have three key features: they describe mechanisms, they span multiple levels, and they integrate findings from multiple fields and/or methodologies. A mechanism, according to Craver, is

> a set of entities and activities organized such that they exhibit the phenomenon to be explained. (Craver 2007b, p. 5)

Entities are the parts or relevant *components* of the mechanism; they engage in activities. Activities are the mechanism's causal components. They are productive behaviors, causal interactions, omissions, preventions, etc. Though this reference to production is likely inspired by Salmon's work Craver explicitly states he does not require any kind of contact or a mark transmission for his activities. But activities are not just temporal sequences or mere correlations either as they are potentially exploitable for manipulation and control. In a mechanism, entities and activities are *organized spatially, temporally, causally, and hierarchically* such that they exhibit the explanandum phenomenon.

Craver pictures this view as shown in figure 3.1. He refers to the phenomenon, behavior, or property a mechanism explains as ψ and the mechanism as a whole as S. At the bottom, circles represent the mechanism's component entities (Xs) and arrows represent their activities (ϕ-ing). According to Craver, "S's ψ-ing is explained by the organization of entities $\{X_1, X_2, ..., X_m\}$ and activities $\{\phi_1, \phi_2, ..., \phi_n\}$." (Craver 2007b, p. 7)[17] The mechanism's lower-level components are *organized* such that they "make up" (Craver 2007b, p. 170) something that is greater than the mere sum of its parts. What Craver means by organization in this context is that "the parts have spatial (location, size, shape, and motion), temporal (order, rate, and duration), and active (for example, feedback) relations with one another by which they work together to do something." (Craver 2007b, p. 189) The interlevel making-up relation Craver talks about here is *constitutive* in nature rather than

---

17 In "Explaining the Brain" Craver sometimes makes it sound as though activities are by themselves components. However, he is very explicit elsewhere (Craver 2007a) that activities are *what entities do*.

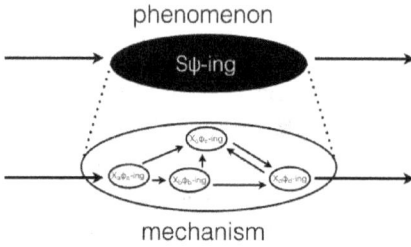

Fig. 3.1: A phenomenon and its mechanism.
See Craver (2007b, p. 7) for a similar illustration.

causal etiological. This marks an important difference to Glennan's view: Craver's mechanistic explanations do not simply list a number of antecedent causes but focus instead on component entities and activities and how they are organized.

On Craver's view mechanistic explanation is an inherently interlevel affair: a mechanism's behavior is on a higher level than the mechanism's component entities and activities. A mechanistic explanation relates these levels; indeed, it can relate multiple levels as components may themselves unfold into lower-level mechanisms. The way Craver views them, *levels* of mechanisms are not in any sense universal or "monolithic divisions in the structure of the world" (Craver 2007b, p. 190). They are always relative to a given mechanism. They are identified within a given compositional hierarchy only. Thus, there is no unified answer to question how many levels there are or when two items are at the same level. "In the mechanistic view, what places two items at the same mechanistic level is that they are in the same mechanism, and neither is a component of the other." (Craver 2007b, p.195, also see p.192, fn11)[18]

Taken together, the above yields the following definition of *mechanistic explanation*:

(ME) A mechanism is constituted by a set of relevant entities and activities organized such that they exhibit the phenomenon to be explained. The mechanism's behavior as a whole (S's ψ-ing) is explained by the orchestrated (temporally organized) activities of (spatially organized) component entities (Xs' ϕ-ings) in the mechanism. Mechanistic explanations are inherently interlevel; S's ψ-ing is at a higher mechanistic level than the Xs' ϕ-ings.

---

18 There is much more to be said on the issue of levels. For current purposes this shall suffice, though. Craver discusses mechanistic levels extensively in chapter 5 of his book. I will come back to this topic in section 3.7 below as well as in chapter 13.

Now if the explaining is done by all of the mechanism's component parts (including their properties, activities, and organizational features), the obvious follow-up question is what qualifies as a component. Craver argues that "[l]evels of mechanism are levels of composition, but the composition relation is not, at base, spatial or material." (Craver 2007b, pp. 188–189) Mere spatial containment is therefore insufficient for mechanistic componency. Rather, mechanistic components must be *constitutively relevant*, viz. relevant for the mechanism's operation, relevant for the organized set of entities and activities to exhibit the explanandum phenomenon. But how do we know whether any given part of a mechanism is relevant in this sense? Craver offers the following working account of *constitutive relevance* (CR):

(CR) a component is relevant to the behavior of a mechanism as a whole when one can wiggle the behavior of the whole by wiggling the behavior of the component *and* one can wiggle the behavior of the component by wiggling the behavior as a whole. The two are related as part to whole and they are *mutually manipulable*. More formally: (i) X is part of S; (ii) in the conditions relevant to the request for explanation there is some change to X's ϕ-ing that changes S's ψ-ing; and (iii) in the conditions relevant to the request for explanation there is some change to S's ψ-ing that changes X's ϕ-ing. (Craver 2007b, p. 153)

The principle idea is fairly straightforward. Assuming we have established a part-whole relationship, if we can *mutually manipulate* the behavior of mechanism as a whole and the activity of a given part of the mechanism this is sufficient to conclude that this part is *constitutively relevant* to the mechanism's behavior; it is a *component* in the mechanism under investigation. Likewise, it is sufficient to show that we cannot manipulate S's ψ-ing by interfering with X's ϕ-ing *and* we cannot manipulate X's ϕ-ing by interfering with S's ψ-ing to conclude that X's ϕ-ing is *not* constitutively relevant to S's ψ-ing (cf. Craver 2007a). This yields the so-called *mutual manipulability criterion* (MM) that functions as a sufficient condition for interlevel constitutive relevance in mechanisms:

(MM) A part is constitutively relevant, viz. a component, in a mechanism if one can change the behavior of the mechanism as a whole by intervening to change the behavior of the component *and* one can change the behavior of the component by intervening to change the behavior of the mechanism as a whole.[19]

---

[19] Craver states that "a part is a component in a mechanism if one can change the behavior of the mechanism as a whole by intervening to change the component *and* one can change the behavior

The mutual manipulability relation is a *symmetrical* one. It is only informative for cases where there is manipulability or non-manipulability in both directions (top-down as well as bottom-up) and does not say anything about intermediate cases where there is manipulability in one direction only (cf. Craver 2007a, p. 17).

As we have seen in chapter 2, finding mutual manipulability relations is indeed a common strategy in scientific practice. It is also widely agreed among contemporary philosophers of science that the norms of constitutive relevance (MM) reflects are implicit in many experimental designs (e.g. Bechtel & Richardson 1993; Bechtel 2008; Craver 2002, 2007b,a; Craver & Bechtel 2007; Fazekas & Kertész 2011; Kästner 2011, 2013; Krickel 2014; Leuridan 2012). Thus, it seems appropriate to think of scientists investigating, say, the neural basis of language processing as manipulating S's ψ-ing (the language processing) or X's ϕ-ing (a neural process believed to be a component in the mechanism for S's ψ-ing) and measuring the changes this brings about in X's ϕ-ing or S's ψ-ing, respectively. Since manipulations can be either *inhibitory* or *excitatory* in nature, Craver (2007b) specifies four different types of interlevel experiments:

(I) **Inference Experiments** are bottom-up inhibitory, i.e. one manipulates X to inhibit its ϕ-ing and observes the effects on S's ψ-ing.
(S) **Stimulation Experiments** are bottom-up excitatory, i.e. one manipulates X to elicit its ϕ-ing and observes the effects on S's ψ-ing.
(A) **Activation Experiments** are top-down excitatory, i.e. one manipulates S to elicit its ψ-ing and observe effects on X's ϕ-ing.
(D) **Deprivation Experiments** are top-down inhibitory, i.e. one manipulates S to inhibit its ψ-ing and observes the effects on X's ϕ-ing.

As any type of experiment may confront complications that do not allow for straightforward conclusions from the observations being made, systematically combining different designs and methodologies is vital in empirical practice. On aspect of this systematic combination is captured by (MM): constitutive relevance can be inferred where bottom-up and top-down studies yield complementary results.

The perhaps most problematic thing about Craver's account is the constitutive interelvel relation between a mechanism and its components. Although the basic idea seems rather intuitive, it has been subject to much debate (see chapters 6

---

of the component by intervening to change the behavior of the mechanism as a whole." (Craver 2007b, p. 141) I take it that he intends the manipulabilty relations to hold between behaviors to infer a constitutive relation between entities. Bottom-up interventions therefore target a component to change the *behavior of the component* just as top-down interventions change the behavior of the mechanism as a whole.

and 7). For now, let me only briefly clarify a few things. First, Craver explicitly states that the mechanism-component relation is *non-causal* (see Craver 2007b, p. 146, fn26 and p. 153, fn33 as well as Craver & Bechtel 2007; Craver 2007a). This is important since parts and wholes are not wholly distinct and thus cannot be causally related—at least not on some standard conception of causation. Second, the mutual manipulability relation holds between a mechanism's behavior as a whole and the activity of *one* component, not the entire set of activities and entities that compose the mechanism. Third, although interlevel mechanistic relations are not supposed to be causal this is not to say there is no causation in mechanisms. Within a mechanistic level there are, quite obviously, causal relationships (activities include causal interactions). Finally, and getting a little ahead of myself, the relation between the collection of organized entities and activities that constitute a mechanism is plausibly interpreted as a *supervenience* relation. But this will be a matter to consider in more detail later (in chapter 4.2).[20]

All in all, though the core idea remains, Craver-style mechanistic explanations have come a long way from MDC-style mechanistic explanations. While MDC's definition of mechanisms made reference to start-up and termination conditions, Craver drops this from his definition entirely. In places, he still presents a mechanism's behavior as producing a certain output in response to a certain input. For the most part, however, he keeps his view even more general than that by simply talking about the "behavior" of a mechanism. MDC's talk about "productivity" is mostly replaced with entities and activities as "bringing about" a certain phenomenon which is arguably closer to scientists' way of talking. While this may merely be a change in wording, the notion of production Craver employs seems to be importantly different from that used by MDC. MDC's emphasis on *productive continuity* and the *production of regular changes* seems to be somewhat close to Salmon-style mark transmission and Glennan-style interactions. Their "productivity" seems to suggest that there is a causal process linking some input to a certain output. Craver's "production", on the other hand, seems to capture more of a general *responsibility for* or *relevance to* the occurrence of the explanandum phenomenon. This kind of production can also be an *interlevel* relation: mechanisms underlie their phenomena. In a way then, Craver's use of "production" may correspond to MDC's use of "making up".[21] Craver also uses the notion of activities in a much more general sense than did MDC. He spells out his account of levels of mechanisms in much more detail and supplements it with an account of interlevel

---

20 I explain the concept of supervenience in chapter 4.2.1.
21 For more on relevance relations in mechanistic explanations see especially chapters 7 and 11. For more on "production" and "making up" see chapter 13.4.

constitutive relevance. The criterion he offers to identify a constitutive relevance relation, (MM), is inspired by and aligns with empirical research practices. Yet, as we will see in part II, it is not without problems.

Also in line with empirical research is the expressed relativity of mechanisms. The decomposition of a mechanism into parts always depends on the explanandum phenomenon and/or on the perspective investigators take towards a phenomenon and/or the tools available for studying it. (I will return to this in chapter 13.) This is a characteristic not only shared with MDC but also with Glennan. The same holds for accommodating for exceptions in mechanistic explanations. Like Glennan and MDC, Craver makes it quite clear that he does not intend universal exceptionless explanations. He goes even further than that in stating that it "is not required that all instances of ψ-ing be explained by the same underlying mechanism. What matters is that each instance of ψ-ing is explained by a set of components that are relevant to ψ in that particular mechanistic context." (Craver 2007b, p. 160) This reflects the idea that phenomena can be *multiply realized*, viz. realized by different mechanisms or substrates—an assumption that has become widely accepted in contemporary neuroscience as well as cognitive science more generally. Note that although there might be multiple mechanisms realizing the same phenomenon in different circumstances, a mechanism for a given phenomenon is supposed to produce this phenomenon regularly in the absence of interferences. Craver does not emphasize this point as much as MDC but he still holds onto it.

Another feature worth noting is that for Craver what does the explaining is the mechanism *as such*. It is the mechanism in the world, the thing out there, that *is* the explanation. This *ontic* view contrasts quite sharply with Glennan's view that mechanistic *models* are explanatory.[22] Yet, Craver also shares an important feature with Glennan: both import Woodward's (2002; 2003) notions of relevance and interventions. This takes us to the next section.

---

22 There is actually quite some disagreement on what precisely the role of mechanisms is and what is explanatory about them. While Craver's ontic conception probably marks one extreme of a continuum, William Bechtel's view of (mechanistic) explanations as "epistemic activities performed by scientists" (Bechtel 2008, p. 18) marks the other (see also Bechtel & Abrahamsen 2005; Illari 2013; Kästner & Haueis 2017). Personally, I am not committed to an ontic view, but neither am I committed to the view that mechanistic explanations are *just* epistemic tools. I believe that scientists ultimately aim to figure out what is really the case but understanding can be gained by models invoking abstract or fictional entities and black boxes, too. Given our epistemic limitations, I doubt we can ever conclusively test whether any explanation actually gets to the bottom of things. My aim is to understand how scientists come up with explanations of cognitive phenomena, independently of how we evaluate them.

3.5 The Marriage with Interventionism

So far I have been presenting Craver's view by talking about manipulations and relevance in a non-technical sense. However, Craver makes it quite clear that he embraces an interventionist view (Woodward 2002, 2003; Woodward & Hitchcock 2003) and the associated concepts of *invariance, production, manipulability, relevance,* and *intervention* (Craver 2007b, ch. 3).[23] When talking about interlevel manipulations, for instance, what Craver has in mind are indeed Woodwardian interventions. And, given what we see in scientific practice, this has some *prima facie* plausibility. Especially since both Woodward and Craver lean heavily on the gold standards of experimental design to characterize the manipulative practices they are interested in.

However, adopting Woodwardian interventions turns out to be problematic in the context of a Craver-style account of mechanisms. I will discuss this issue in detail in chapter 6 but the gist is this: Craver characterizes the interlevel constitutive relevance relations in a mechanism as non-causal part-whole relations. However, the notion of relevance he invokes resembles Woodward's (2003) conception of relevance in terms of difference making. But Woodward's interventionist view is tailored to identifying causal relations while Craver's (MM) is designed to find constitutive ones. And constitutive relations cannot be causal: parts cannot cause their wholes or vice versa. Besides causal relations will not give us mutual manipulability as causes bring about their effects but not vice versa. However, there is also a rather unproblematic sense in which Woodwardian interventions can be used in mechanisms: to investigate and characterize the causal relations among mechanistic components at a single mechanistic level.

3.6 Struggling with Reductionism?

The account of mechanistic explanation Craver presents is a multi-level account. But the idea that mechanisms may *unfold all they way down* from macro-level behavior to molecular processes has called reductionists to the arena. The arguably most radical proposal in this context is John Bickle's *ruthless reducutionism* (Bickle 2006; Silva et al. 2013).[24] Though Bickle conceives of levels as absolute rather

---

23 I will discuss Woodward's view in chapter 4. Though the account Woodward presents in "Making Things Happen" (2003) is disentangled from talk about mechanisms, he clearly states elsewhere that he conceives of explanations as mechanistic in character (e.g. Woodward 2002, 2008a, 2013).

24 Note that Bickle's recent view goes beyond what he defended previously (Bickle 1998, 2003).

than relative to a mechanism, the principle disagreement between mechanists and reductionists does not hinge on their conceptions of levels. Rather, it is about the role that higher-level phenomena play in the explanatory enterprise. According to ruthless reductionism, only complete molecular-level descriptions can adequately explain cognitive phenomena. Higher-level phenomena are themselves denied explanatory relevance. For the ruthless reductionist, research into higher-level phenomena as conducted by psychologists and cognitive neuroscientists is a mere precursor of cellular and molecular neuroscience research, a heuristic tool at best.

The vices and virtues of a reductionistic account of scientific explanation, ruthless or not, are at hand. Being a reductionist, one does not confront worries of causal overdetermination (see chapter 4.2.1). Neither does one have to bother with a multi-layered ontology or such things as interlevel causal relations. All that eventually matters are physical properties of items at the molecular level. The obvious downside is that spelling out all the "gory details" (Kitcher 1984) of the cellular and molecular processes underlying a cognitive phenomenon seems pointless. For even if we mange to do this, the complexity of our explanations will be practically intractable (e.g. Dennett 1991). Besides, even if one could tell the full causal story of, say, remembering a word list in terms of ion currents, protein bindings, perhaps gene expression, understanding these molecular processes seems insufficient to understand a memory's role in the context of an overall cognitive system. There is no straightforward way to read off learning and forgetting curves so prominently reported by experimental psychologists (e.g. Goldstein 2004) from stories about gene expression and electrophysiological properties of nerve cells. It is precisely this link between cellular and molecular processes within biological brains and phenomena observed in the psychological domain is what cognitive scientific research aims to investigate. Hence, ruthlessly reductionistic explanations seem to be missing the point when it comes to understanding explanations in cognitive neuroscience: they are unable to link the explanans (cognitive phenomenon) to its implementing mechanism.

This is where mechanistic accounts in a MDC or Craver fashion come to the rescue. They tell us how higher-level (cognitive, psychological) phenomena are linked to and explained by their lower level (neural) mechanisms.[25] Already MDC argue that

---

[25] As said, mechanistic levels are not general but always relative to a given mechanism (and this holds true not only for individual levels but for entire mechanistic hierarchies). Yet, I will continue to talk about cognitive processes as "at a higher level" and neural processes as "at a lower level". I think it is uncontroversial to think of cognitive processes as somewhere further up in the mechanistic hierarchy than neural ones.

higher-level entities and activities are [...] essential to the intelligibility of those at lower levels, just as much as those at lower levels are essential for understanding those at higher levels. It is the integration of different levels into productive relations that renders the phenomenon intelligible and thereby explains it. (MDC, p. 23)

The process of integration proceeds by recognizing how different theories mutually constrain one another depending on the evidence they are based on (see also Craver & Darden 2005, p. 241). It connects different levels, even different fields of research to yield an "explanatory *mosaic*" (Craver 2007b). Once the mosaic is constructed we can look at it from a distance to see the "bigger picture"; but we can also take a closer look at some of its components and access the details of implementing mechanisms. Since theories at different levels all contribute explanatorily information to this mosaic, we cannot simply reduce theories at one level to theories at another. Besides, higher level processes are explanatorily relevant *because* they are causally relevant (Craver 2007b, p. 198) (within any given level, of course).

In "Explaining the Brain", Craver argues that a mechanistic explanation is reached only once the mosaic is complete, i.e. when the mechanism sketch is fully filled in.[26] If this is true, it seems that the only difference between mechanistic explanation and even as radical a reductionist position as Bickle's is what explanatory value is granted higher and intermediate level descriptions: while reductionists discard them into the explanatory trash can as mere heuristics, mechanists keep them as explanatorily valuable pieces of their mosaic.[27] Against this background, some may conceive of mechanistic explanations as reductive in character. Others will object to this by pointing out that (i) mechanistic explanations ascribe explanatory relevance to higher levels and (ii) mechanistic explanations do not necessarily get *better* if more details are included (Craver & Kaplan 2016; Boone & Piccinini 2016). But whether this actually contradicts any kind of reductionist reading of mechanistic explanations will crucially depend on the precise notion of "reduction" we employ.[28]

---

26 Craver basically shares the distinction between mechanism sketch and mechanism schema advocated by MDC. A mechanism sketch contains missing pieces and filler terms that function as black boxes. A mechanism schema is an abstract description of a mechanism that can be filled in. While MDC seem to attribute explanatory power to such abstract descriptions, Craver takes only the actual mechanism to be explanatory (see also Craver 2006).
27 At least, as long as they are not just fictional stories postulating non-existent entities; but this is a different story (cf. Craver 2006).
28 I will get back to this issue in chapter 13.5.

3.7 Little Ailments

Contemporary philosophy of science largely accepts that interlevel studies and mutual manipulability capture some important aspects of scientific practice. Against this background, Craver's account has been celebrated as particularly empirically adequate. Yet, some notorious problems remain for the mechanistic account. For instance, Craver suggests that identifying both intralevel causal and interlevel constitutive relations is based on Woodwardian interventions. Whereas constitutive relevance is tested by interlevel studies, causal relevance is tested by same-level studies (Craver 2007a, p. 12). The latter is unproblematic, but the former raises a puzzle (see chapter 6).

A related problem concerns (MM). Craver suggests it as a sufficient but not a necessary condition for mechanistic constitutive relevance. Likewise, if there is no manipulability in either direction this indicates there is no constitutive relevance. However, this criterion leaves open what to conclude about the interlevel relation in intermediate cases in which only one half of (MM) is satisfied. Besides, the status of (MM) has been debated given the limits of empirical research (e.g. Leuridan 2012; Mc Manus 2012). Even those who accept (MM) will have to acknowledge that much of what we see in scientific practice is not actually as clean an application of (MM) as it seems. In fact, scientists often carry out separate manipulation experiments in either direction. As a result, these studies very much resemble experiments assessing unidirectional causal rather than constitutive relevance relations; mutual manipulability relations are only identified once the evidence from different experiments gets combined (see chapter 7.2).

Another difficulty connected to (MM) is that the nature of the mechanistic constitution relation is unclear. Recent attempts to spell out this relation (e.g. Harbecke 2010, 2015; Couch 2011) have mostly been focusing on its exact metaphysical properties. However, as Craver leaves the metaphysical underpinnings of his account largely undiscussed these suggestions are hard to evaluate. We may ignore the fact that we do not know what exactly mechanistic constitution is supposed to be for the time being. But even so, (MM) presents a pressing worry. In order to apply it, we need to manipulate mechanistic components; but in order to do that, we must know (or at least guess) what the components are and how to target them using experimental manipulations. Also, once we manage to mutually manipulate two processes, we must know which is the mechanism's behavior and which is the behavior of a component. Without this, we cannot tell in which direction the constitutive relation is supposed to hold. For if A and B are mutually manipulable, we do not know if A constitutes B or vice versa. That is to say that in order to identify constitutive relations, we must be able to decompose a phenomenon and localize its components at different levels before we can even apply (MM). However, the role of localization and decomposition is not uncontroversial. Chemero and Silberstein (2013) for instance discuss cases from *systems neuroscience* to illustrate that not all explanations rely on these strategies. Insofar as the mechanistic view does, they argue, it cannot account for all the different kinds of explanations in science.[29]

Finally, the radical relativity of mechanistic levels raises issues with integrating multi-level explanations into a mosaic. This is because once a mechanism's components get analyzed into mechanisms themselves we do not have, according to Craver, any means of saying that their components are on the same level. Suppose we have a mechanism M with two components C_1 and C_2. Now we find that C_1 and C_2 are actually themselves mechanisms with two components each. C_1 has components SC_a and SC_b, C_2 has components SC_c and SC_d. Since mechanistic levels are only relative to any given mechanism, we cannot say that SC_b and SC_c are on the same level. For they are components in different mechanisms (C_1 and C_2) even though C_1 and C_2 are both components in a higher-level mechanism (M). If this analysis is correct another, possibly more serious, problem follows. Since there can only be causal interaction within any given mechanistic level, a puzzle is raised with respect to how the components C_1 and C_2 in the original higher-level mechanism M can causally interact. If the subcomponents SC_a, SC_b, SC_c and SC_d that are constitutive of C_1's and C_2's causal interaction cannot be said to be on the same level, there cannot be causal interaction between the components of the higher-level mechanism either. For if we cannot link any of the components of C_1 to any of the components of C_2, it seems that we cannot link C_1 and C_2 at all (cf. Fazekas & Kertész 2011). But the problem gets even worse. For if causation can indeed only be an intralevel affair, we may turn the argument around:[30] whenever we identify a causal relation between two entities (say, SC_a and SC_c) we can infer that they must be on the same level. For only entities on the same level may be causally related. If this is correct, mechanists face a dilemma: either they cannot have intralevel causal relations (Fazekas & Kertész's problem) or mechanistic levels

---

29 This conclusion is controversial, too. For instance, Kaplan (2011) argues that at least some computational models in fact describe mechanisms. However, Kaplan & Craver (2011) argue that dynamical systems are not mechanisms (see also Dupré (2013)) and that they are only explanatory if they appropriately map onto a mechanistic model. By contrast Bechtel (2011) holds that at least some dynamical systems qualify as mechanistic and Kaplan and Bechtel (2011) argue that dynamical models and mechanistic explanations are complementary rather than in opposition. Woodward (2013) takes this issue to be primarily about terminology. I agree insofar as what counts as a mechanism will naturally depend on how restrictive a notion of mechanism one defends.
30 Thanks to Anna Welpinghus for pointing this out.

cannot actually be as radically relative as Craver describes them. Perhaps they are not. I will return to this issue in chapter 13.

3.8 Upshot

We have now traced the most crucial steps in the life of mechanistic explanations until today. Craver's (2007b) account is probably the most widely adopted view when it comes to mechanistic explanations in the sciences. It builds on Salmon's C-M model and MDC's view, imports ideas from Glennan (1996; 2002) and heavily leans on some of Woodward's (2002; 2003) ideas. While various alternative suggestions have also been made (see Illari & Williamson (2012)) I will work with a Craver-style mechanistic view in what follows. I will adopt the view that mechanisms are constitutive in nature and consist of organized entities and activities that are somehow responsible for the explanandum phenomenon and the relevant components of which are identified by applying (MM).[31]

Since Craver's view successfully captures the interlevel character of much of scientific practice it is especially apt to fit empirical sciences that relate different domains such as cognitive neuroscience. Explanations in this field are paradigmatically mechanistic. They span multiple levels as they relate processes in the brain to cognitive phenomena. They describe neural mechanisms to explain these phenomena, and they integrate findings obtained using multiple different experiments and/or methodologies. This makes mechanistic explanations a very promising framework when trying to understand how explanations in cognitive neuroscience work.

However, it is precisely in this interlevel context (when relating cognitive phenomena and neural processes) that Craver's view faces problems; some of these have been mentioned already. Before discussing them in more detail in part II, we need to take a closer look at Woodward's *interventionism*. This will be the focus of the next chapter.

---

[31] For a recent discussion of causal and constitutive accounts of mechanisms see Kaiser & Krickel (2016).

4 The Interventionist View

Summary

Woodward's (2003) celebrated *interventionist account of causation* is currently one of the most popular theories of causation and scientific (causal) explanation. The basic intuition behind this view is that causation is a matter of *difference making*. According to Woodward, manipulability reveals causal relations and scientific explanations are based on the manipulability relations scientists identify. This chapter outlines Woodward's motivations, his view, and some refinements developed in response to challenges.

4.1 Interventionism Outlined

As we have seen in the chapter 3.1, Salmon's C-M model presents significant progress over logical positivism in the 1980s. His emphasis on *causal processes* in the context of providing scientific explanations was pioneering for philosophy of science. Yet, Salmon's own account remains rather unsatisfying. It requires physical contact for processes to be causal and lacks a workable account of relevance. James Woodward aims to fix these issues by devising his celebrated *interventionist* or *manipulationist* account of causation, *interventionism* for short. More than that, Woodward promises his account to have "adequate epistemological underpinnings" (Woodward 2003, p. 23), viz. to be accompanied by a plausible epistemological story that aligns with actual research practices.

Woodward's interventionism is a counterfactual theory of causation and scientific (causal) explanation according to which causation is a matter of *difference making*.[1] The interventionist conception of explanation is nicely captured in the following quote:

> we think of [causal explanation] as embodying a *what-if-things-had-been-different* conception of explanation: we explain an outcome by identifying conditions under which the explanandum-outcome would have been different, that is, information about changes that might be used to manipulate or control the outcome. More generally, successful causal explanations consists in the exhibition of patterns of dependency (as expressed by interventionist counterfactuals) between the factors cited in the explanans and the explanandum—factors

---

1 Note that interventionists equate scientific explanations with causal explanations. Causal explanations, according to the interventionist, cite causal relations. Once we uncover causal relations, we therefore discover explanations. Thus, there is no sharp contrast between causal and explanatory relations for the interventionist.

DOI 10.1515/9783110530940-004

that are such that changes in them produced by interventions are systematically associated with changes in the explanandum outcome. (Woodward 2008a, p. 228)

The *what-if-things-had-been-different* character of explanations is central to the interventionist enterprise; successful explanations cite generalization that can tell us what will (or would) happen under a range of different circumstances. Explanations, that is, can answer what-if-things-had-been-different questions, *w-questions* for short. Ideally, of course, these explanations they will include all and only relevant factors.

To enter into a more detailed discussion of the interventionist view, some formal work is in order. Inspired by work on causal modelling (Pearl 2000; Spirtes et al. 1993), Woodward conceptualizes events, properties, states of affairs, etc. as *variables* taking different values.[2] Interventionist causal claims relate these variables in "rather than"-structures. Assume, for instance, fire (F) causes a change in thermostat reading (T). Now we can conceptualize John's putting on the fire as intervention I on F that sets the value of F from f_0 (no fire) to f_1 (fire). In turn, the change in thermostat reading can be conceptualized as changing the value of T from t_1 (low temperature) to t_2 (high temperature). T's taking value t_2 rather than t_1 can thus be seen as an effect of F's taking value f_1 rather than f_0. The manipulability of T by intervening into F indicates—assuming certain other conditions are met—that F is a cause of T.

The arguably "natural way" (Woodward 2015, p. 6) to represent such causal relations is in *directed acyclic graphs*. The causal relationship between F and T, for instance, can be visualized as in figure 4.1. A directed acyclic graph consists of a set

$$F \longrightarrow T$$

Fig. 4.1: Visualization of "F causes T" in the interventionist framework.

V of variables (the nodes) standing in causal relationships with one another and a set of *directed* edges connecting them. If and only if there is a directed path along the directed edges from one variable X to another variabel Y in a causal graph G as well as an appropriate manipulability relation then, according to Woodward,

---

2 For the most part I shall talk about variables representing properties here. However, the precise ontological commitments are not important at this point. There are no clear rules as to what exactly a variable can represent and how much information is included in a single variable. This raises a notorious problem of variable choice (see Franklin-Hall 2016a, for more on this).

X should be considered causally relevant to Y (Woodward 2003, pp. 42–45). This yields the following definition of *interventionist causal relevance* (CR$_i$):

(CR$_i$) A variable X is causally relevant to a variable Y with respect to a causal graph G consisting of the variable set V if and only if (i) in G, there is a directed path from X to Y, (ii) there is a possible intervention on X which changes the value or the probability distribution of Y, while (iii) all other variables Z_i in V which are not on the path leading from X to Y remain unaffected, i.e. fixed at some value z_i. (cf. Woodward 2003, pp. 45–61 and Hoffmann-Kolss 2014)

Interventionists exploit the connection of what happens to some variable Y under an appropriate possible intervention on some other variable X to provide necessary and sufficient conditions for X's causing Y. Note that on Woodward's view whether an intervention is "possible" has nothing to do with human agency or human actions. Interventions do not even have to be actually or practically possible for Woodward. Instead, they must be just be *logically* or *conceptually* possible, irrespective of practical constraints, lack of tools, ethical issues, etc.

Now that we know the principle ideas and formal framework Woodward's interventionism is build on, let us take a look at the core definitions:

(M) A necessary and sufficient condition for X to be a (type-level) direct cause of Y with respect to a variable set V is that there be a possible intervention on X that will change Y or the probability distribution of Y when one holds fixed at some value all other variables Z_i in V. A necessary and sufficient condition for X to be a (type-level) *contributing cause* of Y with respect to variable set V is that (i) there be a directed path from X to Y such that each link in this path is a direct causal relationship; that is, a set of variables $Z_1 \ldots Z_n$ such that X is a direct cause of Z_1, which is in turn a direct cause of Z_2, which is a direct cause of $\ldots Z_n$, which is a direct cause of Y, and that (ii) there be some intervention on X that will change Y when all other variables in V that are not on this path are fixed at some value. If there is only one path P from X to Y or if the only alternative path from X to Y besides P contains no intermediate variables (i.e., is direct), then X is a contributing cause of Y as long as there is some intervention on X that will change the value of Y, for some values of the other variables in V. (Woodward 2003, p. 59)[3]

---

3 (M) incorporates an extension of deterministic to stochastic causation. This is why not only the values of variables but also their probability distributions are mentioned. My focus will be on deterministic cases, however.

(IV) *I* is an intervention variable for *X* with respect to *Y* if and only if *I* meets
the following conditions:

I1. *I* causes *X*.

I2. *I* acts as a switch for all the other variables that cause *X*. That is, certain
values of *I* are such that when *I* attains those values, *X* ceases to depend
on the values of other variables that cause *X* and instead depends only
on the value taken by *I*.

I3. Any directed path from *I* to *Y* goes through *X*. That is, *I* does not directly
cause *Y* and is not a cause of any causes of *Y* that are distinct from *X*
except, of course, for those causes of *Y*, if any, that are built into the
I – *X* – *Y* connection itself; that is, except for (a) any causes of *Y* that
are effects of *X* (i.e., variables that are causally between *X* and *Y*) and
(b) any causes of *Y* that are between *I* and *X* and have no effect on *Y*
independently of *X*.

I4. *I* is (statistically) independent of any variable *Z* that causes *Y* and that
is on a directed path that does not go through *X*.

(Woodward 2003, p. 98)

Put very briefly, (M) states that *X* causes *Y* if and only if the value of *Y* will change
under an intervention *I* on *X* with respect to *Y*. An (appropriate) intervention on *X*
with respect to *Y* is defined as *I*'s assuming some value $I = z_i$ that causes the value
taken by *X* while *I* has to be an intervention variable for *X* *with respect to Y*. Where,
as (IV) states, *I* is an intervention variable for *X* with respect to *Y* if and only if (i) *I*
is causally relevant to *X*, (ii) *I* overrides all other (potential) causes of *X*, (iii) *I* is
causally relevant to *Y* through *X* only, and (iv) *I* is independent of any other causes
of *Y*. While (M) presents a manipulability-based metaphysical non-committal notion of causation, (IV) explicates what possible confounders need to be controlled
for (i.e. held fixed) when performing a suitable intervention.[4] Graphically, we can
think of an intervention into *X* as "breaking" all other arrows into this variable
but leaving intact all remaining arrows in the graph under consideration. The fire
example sketched above may thus be represented as shown in figure 4.2; where *I*
represents John's putting on the fire. The use of interventions can be illustrated

---

4 While (M) defines causation in terms of interventions, (IV) defines interventions in terms of
causation. Woodward's account thus is *circular*. Although he readily admits this, Woodward
contends that his view "is not viciously circular" (e.g. Woodward 2003, p. 105). The issue has been
debated in several places (see e.g. Gijsbers & de Bruin 2014) but shall not concern us here any
further. After all, the intimate connection between causation and intervention does not seem to
do interventionism's success any harm. It might even be considered a virtue as these concepts
sometimes appear inseparable both in scientific and everyday contexts.

$$I \longrightarrow F \longrightarrow T$$

Fig. 4.2: Visualization of intervention *I* on *F* with respect to *T*.

considering the well-known case of the barometer. Suppose we observe the reading of the barometer drop. We know that a drop in the barometer reading is usually followed by a storm. We may thus think that the falling barometer reading (*B*) causes the occurrence of the storm (*S*). We can test this hypothesis using an intervention. Say we break the barometer or fix its reading manually; graphically that means we break all arrows into *B*. If this does not affect *S*, we can conclude that *B* is not a cause of *S* and that something else must bring *S* about. Indeed, it's the drop in atmospheric pressure *A* that causes both the occurrence of the storm and the dropping of the barometer reading. Given this causal knowledge, we know that if we were to manipulate *A* we would thereby affect both *S* and *B* (unless, by some other intervention, we were to hold *B* fixed independently).

This appeal to difference making is supposed to capture some salient features of the manipulative nature of experimental science: when trying to understand and explain how a phenomenon comes about or how it works, scientists systematically manipulate it. Be this cars, clocks, toilet flushes, lightbulbs, vending machines, computers, genetic coding, or—as illustrated in chapter 2.3—(human) brains and cognitive processes. Interventionism takes manipulability (at least where it is induced by interventions) to indicate causal relevance. Thus, interventionists hold, there is *no causal difference without a difference in manipulability relations.* Call this the interventionist manifesto. The question "Does *X* cause *Y*?" can thus be equated with the question "Does *Y* change under suitable interventions of *X*?".

Given that interventionists directly infer causal relations from observed intervention-induced manipulability, they must make sure that no confounding factors or covariates are included in the set of variables under consideration; for these would otherwise mistakenly be identified as causes. There are several ways to express this. The most intuitive is perhaps that variables in *V* must not be conceptually related in any way such that every element of *V* can be independently manipulated. Woodward captures this requirement in his *independent fixability criterion* (IF):

(IF) a set of variables *V* satisfies independent fixability of values if and only if for each value it is possible for a variable to take individually, it is possible (that is, "possible" in terms of their assumed definitional, logical, mathematical, or mereological relations or "metaphysically possible") to set the variable to that value via an intervention, concurrently with each of the other variables in *V* also being set to any of its individually possible values by independent interventions. (Woodward 2008a, p. 12)

A more graph-theoretic way to express this is to require that a causal graph *G* must satisfy the so-called *Causal Markov condition* (CMC). There are several different versions of this condition but essentially it boils down to this:[5]

(CMC) Every variable must, given its direct causes, be statistically independent of any other variable except its effects.

The Markov condition mirrors some familiar principles of experimental design, such as ideas about *screening off* and *controlling for* potentially interfering factors. Some of these are already captured in Reichenbach's (1956) well-known *common cause principle* (CP):

(CP) Whenever two events *A* and *B* are correlated, either *A* causes *B*, *B* causes *A*, or *A* and *B* are both brought about by a *common cause C* that is distinct from *A* and *B*.

However, (CMC) is stronger than (CP); it might even be seen as a generalization of it. While (CMC) is widely accepted among philosophers of science (e.g. Hausman & Woodward 1999; Strevens 2008; Eronen 2010; Baumgartner 2010; Hoffmann-Kolss 2014), it is not uncontroversial. Cartwright (2002) for instance offers a counterargument. Even Woodward himself is—surprisingly given his appeal to experimental practice—hesitant to embrace (CMC) and make it part of his interventionist account (cf. Woodward 2003, p. 117). Nevertheless, it has been pointed out that (CMC) follows from (M) and (IV) together with some other plausible assumptions (see Hausman & Woodward 1999; Eronen 2010).

Interventionism not only reflects the manipulative character of scientific practice, it also takes into account that scientists typically subsume the outcomes of their experiments under *generalizations*. (M) mirrors this as it aims to characterize causation between types, not tokens. It is intended, Woodward holds, as an explication of generalizations such as "Smoking causes lung cancer" rather than token claims like "John's smoking caused his lung cancer." In "Making Things Happen" Woodward discusses at length how the interventionist account can also be accommodated to deal with token causation. In the current context, however, we can

---

5 (CMC) is the assumption that a causal system that fulfills certain criteria meets the Markov condition. The Markov condition states that any node in a Bayesian network is conditionally independent of its non-descendants given its parents.

focus on type causation, because the scientific explanations we are interested in are general in character. They explain types of phenomena rather than specific tokens and account for them in terms of conditions that will usually and reliably (regularly) elicit them. Scientific explanations, that is, cite what Woodward calls *invariant change-relating* generalizations.[6] They are change-relating, or contrastive, because they relate changes in some variable X (due to intervention I) to changes in another variable Y (as in the example of John putting on a fire mentioned above). This is captured in (M) which links the claim "X causes Y" to a claim about what would happen to Y if X would be manipulated in a certain way—a change-relating generalization.

For a generalization to describe a causal relationship, it must not only be change-relating but also *invariant* (or stable) under some appropriate (IV)-defined interventions (Woodward 2003, ch. 6). Although Woodward considers this invariance necessary and sufficient for causation, he acknowledges that it may be limited. This is important. For the explanatory generalizations Woodward is after are supposed to capture the relevance relations scientists identify through experiments. Besides, the generalizations scientists work with are quite unlike the strict exceptionless laws figuring in D-N explanations: they are exception-prone and often have a limited scope of application (they apply under appropriate background conditions only). Thus, just as scientific explanations may apply as long as certain background conditions are in place only, Woodward's change-relating generalizations may hold for *some* range of interventions only. If X causes Y then *some*—but not all—interventions I on X with respect to Y will actually induce a change in Y. This is why change-relating generalizations are significantly different from laws of nature: they may have limited application and do not have to hold universally. They can accommodate for *threshold effects* or *switches*. Think, for instance, about heating water. Heating a cup of water for 30 seconds may raise the temperature to 60°C, heating it for 60 seconds may raise the temperature to 100°C. Heating it for 90 seconds may still only raise the temperature to 100°C, however. This is because once it reaches boiling point (a threshold), it is not possible to heat water any further (at least not under normal conditions). Generalizations relating heating time to water temperature will thus only be applicable within certain limits (up until boiling point is reached). By contrast, laws should be universally applicable and not fail to hold once a certain threshold is reached.

Thus, interventionism gets away without having to appeal to laws. As long as a generalization holds (is invariant or *stable*) under some range of manipulations,

---

6 See also chapter 3.

it can be used for explanatory purposes.[7] Moreover, interventionism's appeal to manipulability and relevance relations arguably solves the problem of explanatory relevance that e.g. the D-N and C-M accounts of scientific explanations face. If, for instance, we intervene into John's taking birth control pills we will soon realize that it does not make a difference (i.e. is not relevant) to his non-pregnancy. John fails to get pregnant with as well as without taking the pills. In the case of John's wife, however, intervening into her taking birth control pills does (at least in certain circumstances) make a difference to her non-pregnancy; it thus is relevant.[8] Similarly, interventionists can handle causation at a distance: intervening into gravitational fields makes a difference to the movements of the planets. Thus, gravitation is causally relevant to the planetary movements irrespective of physical contact or a mark being transmitted.

Overall then, interventionism and its appeal to difference making seem promising to capture salient features of scientific practice as well as meet the challenges leveled against Salmon's C-M view. However, interventionism has recently come under attack from a different camp, viz. debates on mental causation.

4.2 The Issue With Mental Causation

While some have euphorically celebrated that interventionism can solve the problem of mental causation (e.g. List & Menzies 2009; Menzies 2008; Raatikainen 2010; Shapiro 2010; Shapiro & Sober 2007; Woodward 2008a), others have argued it cannot (e.g. Baumgartner 2010, 2013; Hoffmann-Kolss 2014; Kästner 2011; Marcellesi 2011; Romero 2015), not even in light of recent refinements that Woodward offered to his theory (especially Woodward 2008a, 2015). While I do not want to enter into this debate in detail, some developments in this context will be informative for our discussions of explanations in cognitive neuroscience. After all, the mental causation problem shares an important feature with prototypical explanations in cognitive neuroscience: in both cases, we are trying to relate processes at different levels. Therefore, the criticism interventionism has been subjected to in the context of mental causation and the refinements that have been offered will be of interest in subsequent chapters. The current section sketches these developments in somewhat broad strokes.

---

7 While Glennan 2002 used invariant change-relating generalizations to replace strict laws of nature (see chapter 3.3.3), Woodward himself neither affirms nor rejects the existence of strict laws.

8 Interventionism's success in actually solving the problem of explanatory relevance is a matter of debate, however. See Marcellesi (2013) for an argument why the problem remains unsovled.

4.2.1 Kim's Exclusion Argument

Put very briefly, the problem of mental causation concerns the question whether or not mental properties can be causally efficacious. Take two mental properties M and M^* and two physical properties P and P^* that are their *supervenience bases*. There are several different varieties of supervenience (see McLaughlin & Bennett 2011). For the time being the following simple conception will be sufficient:

(SUP) If A supervenes on B then every change in A (the supervening property) will necessarily be accompanied by a change in B (the supervenience base).

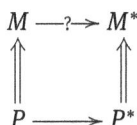

$$M \xrightarrow{\ ?\ } M^*$$
$$\Big\uparrow \qquad\quad \Big\uparrow$$
$$P \longrightarrow P^*$$

Fig. 4.3: The problem of mental causation. Double-tailed arrows indicate supervenience relations.

Now consider the scenario depicted in figure 4.3. We observe both M and P (to use Kim's words, they are both *instantiated*) and we know that there is a purely physical chain of causal processes leading up from P to P^*. When P^* is instantiated, we also find M^* instantiated. Now how come M^* is instantiated?[9] Jeagwon Kim famously puts this problem as follows:

> What is responsible for, and explains, the fact that M^* occurs on this occasion?" [...] there are two seemingly exclusionary answers: (a) "Because M caused M^* to instantiate on this occasion," and (b) "Because P^*, a supervenience base of M^*, is instantiated at this occasion." [...] Given P^* is present on this occasion, M^* would be there no matter what happened before; as M^*'s supervenience base, the instantiation of P^* at time t in and on itself necessitates M^*'s occurrence at t. This would be true even if M^*'s putative cause, M had not occurred [...]. (Kim 2005, pp. 39–40)

According to Kim then, we do not need M to explain the occurrence of M^*. Why is this? Basically, Kim assumes two principles: (i) that there is no overdetermination, and (ii) that the physical is causally closed. (i) gives us reason to suppose we cannot

---

9 "Being instantiated" in this context means that the corresponding property occurs. It is does not indicate a realization relation.

have both M and P^* bring about M^*, this is why we get the two competing answers Kim mentions. His reason to select (b) is stated quite clearly above: it does not even need a cause of P^* for P^* to necessitate M^*. That said, let me add a further note. One might think to save the causal powers of the mental by thinking M could have caused P^*. This, however, is implausible for two reasons. First, we can argue just the same way we did before, i.e. that it does not even need *any* cause of P^* for P^* to necessitate M^*. Second, Kim would not allow for M to cause P^* due to (ii). The physical is causally closed. That is, there is neither mental-to-physical nor physical-to-mental causation. Kim's conclusion then is that the mental is causally pre-empted by the physical—or, more generally, that supervening properties cannot be causally efficacious.[10]

However, this conclusion strikes (not only) philosophers as counterintuitive. After all, it seems that mental properties, events, and processes can exert a causal influence. How else would, say, cognitive behavioral therapy work? Or other psychological interventions? Trying to refute Kim's conclusion, some philosophers have appealed to interventionism. Their efforts focus on demonstrating that mental properties are not causally pre-empted by their physical supervenience bases or, as they may also be called, *realizers*. Interventionism can, they claim, tell a plausible story as to how both supervening properties and their supervenience bases can be causally efficacious.

In order to see how interventionism might deal with Kim's exclusion argument, we first need to clarify what role supervenience can play within this framework. Essentially, if the property represented by some variable M supervenes on the property represented by some other variable P the following should hold: for every value of M, the value of M cannot be changed without being accompanied by a change in the value of P. Put slightly differently that is to say that changes of the value of M necessitate changes in the value of P.

The inevitable complication is already obvious: interventionism equates difference making with causation; now if there is difference making that is not due to causation but some other kind of dependence (such as supervenience), the interventionist has a problem. For, at the very least, she cannot tell apart causal and non-causal dependency relations. However, interventionists have tried to respond to this problem using two different strategies. I will consider the two most prominent replies in turn.

---

10 Note the relevance of Kim's problem for cognitive neuroscience. If we conduct interlevel manipulations, basically what we do is manipulate at the M-level to assess effects at the P-level or vice versa. There will be more on this issue in chapter 6.

4.2.2 Realization Independence

It is generally agreed that we cannot put supervening variables in the same interventionist graph (e.g. Eronen 2010; Hausman & Woodward 1999; Hoffmann-Kolss 2014; Shapiro & Sober 2007; Woodward 2008a).[11] A straightforward consequence would be to *not* use interventionism in cases where supervenience is involved. However, this would limit the applicability of interventionism tremendously; and it would prohibit using interventionism in many places where is particularly intuitively appealing—including scientific explanations in psychology and cognitive neuroscience.

One possibility to work with interventionism despite the presence of supervenience relations is to re-describe graphs with supervening variables as two different graphs. For instance, if X supervenes on R, and we can carry out interventions on both X and R with respect to Y, we could draw the two graphs depicted in figure 4.4. The question remains, however, whether X is causally pre-empted by R (as is M by

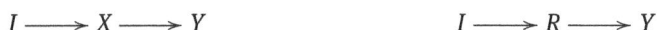

$$I \longrightarrow X \longrightarrow Y \qquad\qquad I \longrightarrow R \longrightarrow Y$$

Fig. 4.4: Two graphs depicting manipulability of Y through intervention into X and R, respectively.

P according to Kim's exclusion argument). Or at least, under which conditions we should prefer talking about X or R, respectively, as causing Y. Kim's conclusion that P pre-empts M, the interventionists' argument goes, was based on the assumption that the physical is somehow ontologically prior to the mental. This assumption is not shared by interventionism. And after all, this would not be faithful to what we find in the empirical sciences such as psychology and cognitive neuroscience where higher-level causal claims are often used for explanatory purposes.

To escape the exclusion problem, however, the interventionist still needs to tell us something about when higher-level causal claims are preferable to lower-level ones. When, that is, should we cite X as a cause of Y rather than R? Woodward (Woodward 2008a) argues that higher-level causal claims are more adequate then lower level causal claims if and only if they express a *realization independent dependency relationship*. While Woodward himself does not offer a precise definition of his notion of a realization independent dependency relationship (RID), Hoffmann-Kolss (2014) reconstructs Woodward's view in the following way:

---

11 By "supervening variables" I mean variables representing properties standing in a supervenience relationship.

(RID) There is a realization independent dependency relationship between two variables X and Y belonging to some causal graph G (short: X is realization independent with respect to Y) if and only if (i) X is classified as causally relevant to Y according to the interventionist account and (ii) any intervention changing the value of a variable representing X's realizers, but not the value of X itself, while keeping the values of all other variables in G fixed, does not change the value or the probability distribution of Y. (Hoffmann-Kolss 2014, p. 58)

The overall idea thus seems to be that realization independence holds when a supervening variable remains unchanged under a number of changes to its supervenience base. Intuitively, we may think about this as a parsimony principle in the context of multiple realization: whenever a variable has multiple realizers we can cite the multiply realized variable in our causal (explanatory) generalization rather than the different realizers.

However, there are at least three problems with (RID). First, it yields different results in different contexts such that we do not know whether higher-level variables or their lower-level realizers should be attributed causal relevance for a given effect; for illustration see Hoffmann-Kolss (2014). Interventionism thus faces an underdetermination problem. Second, some of the graphs (RID) produces violate (CMC). Put very crudely, this is because (RID) introduces realizer variables into causal graphs. However, such non-causally dependent variables cannot be included in interventionist graphs according to (CMC). If we want to get around this problem, we could choose to include *either* realizers *or* supervening variables. However, we lack a clear criterion that tells us which of the variables to include. Third, (RID) requires us to know about supervenience or realizer relations to start with. It only tells us something about the relation between X and Y *provided that* we know which other variables represent realizers of X.[12] As long as we are merely trying to relate mental or cognitive phenomena to neural processes, this last point may not be a major problem. For we might simply build on the general assumption that mental phenomena are implemented by processes int the brain. But even so, note that this problem may arise in different contexts where where we do not know which variables represent the realizers and which the supervening phenomena. Taken together, and without discussing the very details of any of these challenges, it is quite evident that (RID) is not a convincing solution to the problem of mental causation. So let us turn to the second candidate.

---

12 As we will see in the next section, these problems essentially remain unsolved by Woodward's (M*)-(IV*)-interventionism.

4.2.3 Interventionism Revised: (M*) & (IV*)

In 2011, Woodward set out to devise another reply to the causal exclusion problem. This time, he focused on demonstrating how interventionism is supposed to handle graphs containing different kinds of dependency relations, such as causal and supervenience relations as in Kim's famous diagram (see figure 4.3). The challenge for Woodward is, again, to combine a definition of causality in terms of difference making with difference making that is not due to causation. Problematically, such non-causal difference making should normally be controlled for according to the interventionist view. This means Woodward now has to handle "badly designed" experiments. He has to tell a plausible story of how interventionists can deal with cases where interventions on some variables will "automatically" change other variables without there being a causal connection between the two.

(IV) as it stands, Woodward admits, does not give us a handle on Kim's graphs. After all, "ordinary causal graphs are very different from diagrams like Kim's" (Woodward 2008a, p. 19); (IV) and (M) "simply do not transfer to contexts in which non-causal relations of dependence are present" (Woodward 2015, p. 326) and (IF) is violated. But the crux is, Woodward admits, that "one can't simply assume that because it is appropriate to control for ordinary confounders in cases in which no non-causal dependency relations are present, it must also be appropriate to control for factors like supervenience bases which do represent non-causal dependency relations." (Woodward 2015, p. 336) In other words: while interventionists certainly do want to control for (i.e. hold fixed) confounding variables, accidental correlates, etc.—and very plausibly so—they do not want to control for supervenience bases—which would be impossible anyway. But (M) and (IV) demand this.

To solve this problem, Woodward presents the following revised definitions of causation and intervention in (M*) and (IV*), respectively:[13]

---

13 The published (2015) version of Woodward's paper does not contain explicit definitions of (M*) and (IV*). The version of (IV*) quoted here is from his 2011 manuscript. Baumgartner (2010) and Gebharter & Baumgartner (2016), along with many others, cite (M*) and (IV*) in this form. In his published paper, Woodward (2015) drops the explicit definitions of (M*) and (IV*) and only introduces them as he responds to Baumgartner (2010). In his reply, Woodward argues that (M*) and (IV*) as put by Baumgartner in fact express what he (Woodward) already intended to state with (M) and (IV). So rather than offering a revision of interventionism in his later writings, Woodward (2015) merely offers a clarification of Woodward (2003). He emphasizes (in line with Woodward 2008) that he never intended interventionism to transfer to cases like Kim's. However, looking at (M*) and (IV*) will be useful to highlight what the problem is with using interventionism in the context of mechanistic explanations (see chapter 6).

(M*) X is a (type-level) cause for of Y with respect to a variable set V if and only if there is a possible (IV*)-defined intervention on X with respect to Y (i.e. that will change Y or the probability distribution of Y) when all other variables in Z_i in V are held fixed, except for (i) those on a causal path from X to Y, and (ii) those related to X in terms of supervenience.

(IV*) I is an intervention variable for X with respect to Y if and only if I meets the following conditions:

I1*. I causes X.

I2*. I acts as a switch for all the other variables that cause X. That is, certain values of I are such that when I attains those values, X ceases to depend on the values of other variables that cause X and instead depends only on the value taken by I.

I3*. Any directed path from I to Y goes through X or through a variable Z that is related to X in terms of supervenience.

I4*. I is (statistically) independent of any variable Z, other than those related to X by supervenience, that causes Y and that is on a directed path that does not go through X.

The key feature of (M*) and (IV*) is that they are relativized to supervening variable sets. Comparing (M*) to (M), two major changes can be noted. First, the definitions of direct and contributing causes have been collapsed to make the definition more accessible. This is uncontroversial; Woodward (2008a) himself does this when recapitulating (M). Second, an exception clause is added to ensure that supervenience bases do not have to be controlled for. This is the purpose of (ii). In (IV*), I3*. and I.4* are both changed to accommodate for changes in the supervenience base of X that occur once X is manipulated. Accordingly, even if a variable representing X's supervenience base is not on the $I - X - Y$ path, changes in this variable will not prohibit I from being an intervention variable on X with respect to Y.

However, as Baumgartner (2010; 2013) has convincingly argued, even (M*)-(IV*)-interventionism "does not pave the way towards an evidence-based resolution of the problem of causal exclusion." (Baumgartner 2013, p. 26). Why is this? Essentially it is because there is an underdetermination problem for interventionism even on the (M*)-(IV*)-account. For we simply cannot, based on manipulability alone, distinguish between different causal hypotheses. Think about Kim's graph (figure 4.3) again. According to Baumgartner (2010) we cannot distinguish between the two causal hypotheses shown in figure 4.5 based on manipulability alone (see also Gebharter (2015)). This is because intervening into M will affect M^* via the changes induced in P and P^* even if M does not cause M^*. Thus, we cannot know,

Fig. 4.5: Baumgartner's (2010) underdetermination problem for Kim-style graphs.

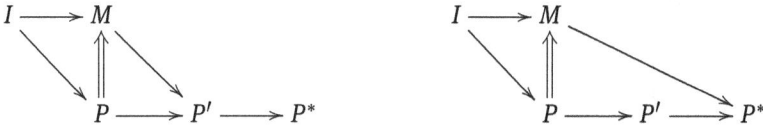

Fig. 4.6: According Baumgartner (2013), the graph on the left is compatible with (M*)-(IV*)-interventionism while the one on the right is not.

by intervention alone, whether or not M causes M^* in *addition* to P causing P^*, M supervening on P, and M^* supervening on P^*.

A related problem comes up when we consider downstream (distant) causal effects of changes in a supervenience base. Consider figure 4.6. While the graph including a causal link from M to P' is compatible with (M*)-(IV*)-interventionism, the graph including a causal link between M and P^* is not. This is because according to (M*)-(IV*)-interventionism, we would have to hold P' fixed on the assessment of the M-P^*-relation. This, however, we cannot do because P' is causally influenced by P which is a supervenience base of M. The conclusion (M*)-(IV*)-interventionism leads to thus is: mental properties may cause physical properties, but only those that are immediate direct effects of changes in the mental property's supervenience base. But this is implausible. Consider the following case: say we learn that the causal chain from P to P^* is not just includes P' but an additional variable P''. Now, all of a sudden, we may no longer think about M causing P' but only P''. This is not only counterintuitive, it also violates the interventionist manifesto: we can still manipulate P' by intervening into M but may not consider P' a cause of M.

4.3 Upshot

Woodward's interventionism undoubtedly has a number of virtues. Its manipulative character and focus on w-questions make it an empirically realistic view of scientific (causal) explanation. This is especially true in the light of the experimental practices I discussed earlier (see chapter 2.3). The reference to invariance captures the intuition that experimental results must generally be reproducible, though within limits. And the concept of invariant change-relating generalizations allows for exceptions and ceteris paribus claims as we find them in actual scientific explanations. Besides, interventionism does not require any kind of physical contact for causation. This makes it attractive and at least potentially applicable to higher-level causal relations, e.g. relations between cognitive processes or those between cognitive and neural processes. As interventionism is primarily an account of type (direct, total) causation rather than token (actual) causation, it mirrors the explanatory practices in science. For typically if scientists aim to explain and understand how a given phenomenon comes about, they aim for a general (type-level) story, not for a story of how this particular token phenomenon came about.[14]

Given these virtues it is not surprising that philosophers have been getting excited about interventionism. Particularly in the context of debates about mental causation and special science explanations (see chapter 3), Woodward's notions have been adopted. However, several problems remain, even in light of a revised (M*)-(IV*)-interventionism. So far, I have only touched upon the problem of mental causation. While some aspects of this discussion will be important later, my focus in the following chapters will be on how interventionism can and cannot be used for (primarily interlevel) explanations in cognitive neuroscience and how it relates to the mechanistic view.

---

14 It is obviously not true that scientists are interested in type-level explanations *only*. Token-level explanations are an important part of science. However, the kinds of explanations we are discussing in this book are predominantly type-level explanations: explanations of how certain phenomena are generally produced.

5 Intermezzo: What's at Stake?

The stage is now set. I have introduced the main characters of the play to come, viz. mechanistic explanations and interventionism. Their quest is to tell an empirically plausible as well as philosophically satisfying story of how scientists explain cognitive phenomena.

In chapter 2, I illustrated how the transition from psychology to cognitive neuroscience reshaped the empirical investigation of cognitive phenomena. Today, scientists typically try to explain cognitive phenomena by reference to neural structures and processes in the nervous system that they hold responsible for producing the phenomena in question. Cognitive neuroscientists study a wide variety of cognitive phenomena including such paradigmatic cases as memory and language processing. And they are using a large pool of methodologies to do so. When they are trying to relate two different domains or *levels*, they often appeal to *interlevel* studies, viz. experiments that manipulate the neural domain to assess effects in the psychological (or cognitive) domain or vice versa. Many of these studies are based on correlational measures of cognitive task performance and brain activity. Although such studies have only become possible once powerful functional imaging techniques were developed, they have been remarkably successful since. They have, quite literally, "brained up" psychology.

The interlevel character of cognitive neuroscience raises some peculiar challenges about relating different levels or domains in the construction of scientific explanations. Given its methodological variety, cognitive neuroscience exhibits characteristics we find in many different special sciences. Therefore, if we can develop an account of scientific explanation that meets the demands of cognitive neuroscience, this account is likely to generalize to other special sciences or, at the very least, function as a paradigmatic illustration for what challenges contemporary philosophy of science faces and how they might be met.

Since the era of logical positivism, philosophy of science has focused on *laws* in the context of scientific explanations. However, I illustrated in chapter 3 that philosophers' attention has shifted towards *causal processes* in the 1980s. Inspired by 17th century mechanical philosophy, new mechanistic accounts have been developed in the late 20th century. The probably most detailed account of mechanistic explanations is provided by Carl Craver (2007b). A major virtue of Craver's approach is that it incorporates the interlevel experiments and spells out how different levels can be related. According to Craver, a (cognitive) phenomenon is explained by the (temporally) organized activities of (spatially) organized entities in the underlying (neural) mechanism. To find a mechanistic explanation for a given phenomenon is to identify the constitutively relevant parts (or components)

DOI 10.1515/9783110530940-005

and how they (causally) work together. This is achieved by applying interventions to identify relationships of manipulability. Put in a nutshell, the idea is that inter-level mutual manipulability indicates a constitutive relation while unidirectional intralelvel manipulability indicates a causal relation.

The mechanistic concepts of manipulability and interventions are imported from James Woodward's (2003) interventionist account of causation. As outlined in chapter 4, interventionism conceptualizes causal relations in terms of difference making. Woodward conceptualizes properties, processes and events as variables taking different values. According to his view, X causes Y if and only if the value of Y will change under an intervention I on X with respect to Y while I has to be an intervention variable for X with respect to Y. I is an intervention variable for X with respect to Y if and only if (i) I is causally relevant to X, (ii) I overrides all other (potential) causes of X, (iii) I is causally relevant to Y through X only, and (iv) I is independent of any other causes of Y (see pages 69–70).

One of interventionism's major benefits certainly is that it reflects some prin-cipal ideas about screening off and controlling for confounders that are familiar from experimental design and causal modeling. Another benefit clearly is that Woodward does not constrain causality to any kind of physical interaction and thus he can handle cases of causation at a distance or causation by omission. Thus Woodward argues that he offers an empirically adequate understanding of special science explanations. However, Woodward's account is not designed to handle interlevel relations and faces—even with amended definitions—severe challenges when forced to handle non-causally dependent variables.

Since Craver builds on Woodward's ideas about manipulability and differ-ence making, contemporary philosophy of science typically treats mechanistic explanations and interventionism as a "package deal". However, they are not as easily married as typically assumed. The major problem is that mechanistic inter-level relations are constitutive in nature, not causal. Yet, mechanists want to use Woodward-style manipulability to assess these relations. Woodward's theory is designed to uncover causal relations, however. Plus it is not really up to handling variables that are non-causally dependent at all. Problematically, variables repre-senting a mechanism and a component (viz. a constitutively relevant part of the mechanism) will be precisely that; for wholes non-causally depend on their parts. Chapter 6 will discuss this problem in more detail and chapter 11 will suggest a strategy to address it.

Viewed against the background of empirical research practice, there are fur-ther challenges for interventionism and mechanistic explanations. For instance, the mechanistic view assumes that we can tease apart interlevel mutual manipula-bility from intralevel unidirectional manipulability based on empirical evidence. However, as chapter 7 will make clear, this is not necessarily the case. Similarly, as chapter 8 will outline, using interventions to assess dependency relations requires us to know what the potential components of a mechanism are in the first place. As it stands, contemporary "mechanisms and interventions"-style philosophy of science nicely cashes out some aspects of scientific practice; but it captures only what is happening somewhere in the middle of the discovery process. Besides, as I will argue in chapter 9, the exclusive focus on interventions is too short-sighted. If we want a fuller understanding of how scientists explain phenomena, we must supplement mechanisms and interventions with something that tells us how we get started in finding mechanisms. As a first step into this direction, I will examine different experimental manipulations scientists use and develop a catalog of experiments in chapters 12 and 14. Along the way, I will briefly discuss an ontologically parsimonious and empirically realistic understanding of interlevel relations in chapter 13.

Now that the protagonists have been introduced, our play will commence. In the second act, we will see how our protagonists struggle with the challenges just outlined. In the final act, we will see how they can eventually be married and can—provided with a new set of tools—succeed in their quest.

Part II: **Puzzles**

"Correlation doesn't imply causation,
but it does waggle its eyebrows suggestively and
gesture furtively while mouthing 'look over there'."

—XKCD

6 The Unsuccessful Marriage

Summary
Craver's (2007b) mechanistic explanations successfully capture the interlevel nature of empirical research in cognitive neuroscience. Without the concepts imported from Woodward's (2003) interventionism, they do not get off the ground, however. While Woodward's interventions do capture some core principles of experimental design, they are not suited for handling mechanistic interlevel relations. This renders the marriage between mechanistic explanations and Woodwardian interventions highly problematic.

6.1 Intra- vs. Interlevel Manipulations

As we have seen in chapter 3, mechanistic explanations appeal to manipulability in an interventionist fashion to cash out intralevel causal as well as interlevel constitutive relationships. The assessment of causal relationships with interventionist tools is obviously unproblematic. After all, this is what interventionism was designed for. But as we shall see, applying interventions across levels is somewhat controversial. The crux is that mechanists employ Woodwardian interventions to assess both interlevel and intralevel relations. However, they contend that interlevel relations are *constitutive* in nature rather than causal.

> Because all of the causal relations are intralevel, there is no need to stretch the notion of causation so that it can accommodate interlevel causal relations. (Craver & Bechtel 2007, p. 562)

As discussed in chapter 3, according to (MM) mutual manipulability between a phenomenon and a component part of its mechanism is sufficient for *interlevel constitutive relevance* (Craver 2007b, pp. 153–156). Note that this constitutive relevance relation holds between an *individual* component of the mechanism and the mechanism that exhibits the phenomenon as a whole. It thus is a relationship between a whole and *one* of its parts rather than a relationship between a whole and its assembled parts (see Craver 2007b, p. 153, fn. 33; Craver 2007a, pp. 15–16).

Since manipulability goes both ways, viz. bottom-up and top-down, the mechanistic constitution is accompanied by a certain *symmetry*. And, importantly, it is supposed to be *non-causal* relation, i.e. if we induce a change in S's ψ-ing by manipulating X's ϕ-ing in a bottom-up study and change X's ϕ-ing by manipulating S's ψ-ing in a top-down study, we should not conclude that X's ϕ-ing caused S's ψ-ing and vice versa. For, within the mechanistic framework, "one ought not to say that things at different levels *causally* interact with one another" (Craver 2007b, p.

DOI 10.1515/9783110530940-006

195; emphasis added). In fact, interlevel mechanistic relations cannot be causal at all. This is because components are related to their mechanisms as parts and whole. And parts cannot, by definition, cause wholes or vice versa. For—at least according to standard conceptions of causation—causes must be wholly distinct from their effects (e.g. Craver & Bechtel 2007).

Yet, mechanists appeal to interventions when they describe bottom-up and top-down studies. Craver (2007b, ch. 3 & 4) even explicitly refers to Woodward's (2003) notion of an *intervention* to spell out (MM). The basic mechanistic suggestion is that interlevel constitutive relations are assessed just like intralevel causal (etiological) ones, only that the intervention and the detection of its effect are applied at different levels.

> In experiments used to test claims about etiological relevance, one intervenes to change the value of a putative cause variable and one detects changes (if any) in the effect variable. Claims about constitutive relevance are tested with interlevel experiments. In interlevel experiments, in contrast, the intervention and detection techniques are applied to different levels of mechanisms. (Craver 2007a, p. 12)

It is certainly true that this conception matches our intuitive reading of bottom-up and top-down studies. However, bottom-up and top-down studies do typically also give us the impression of interlevel causation as we wiggle at one level and observe the effect on another level. Yet, Craver and Bechtel (2007) have argued that such scenarios really need to be analyzed in terms of *mechanistically mediated effects* which they characterize as hybrids of causal and constitutive relations:

> Mechanistically mediated effects are hybrids of constitutive and causal relations in a mechanism, where the constitutive relations are interlevel, and the causal relations are exclusively intralevel. (Craver & Bechtel 2007, p. 547)

That is, in a bottom-up experiment we ought to analyze a change in S's ψ-ing due to manipulation of X's ϕ-ing in terms of an intralevel causal plus an interlevel constitutive relation. The same goes for a change in X's ϕ-ing due to the manipulation of S's ψ-ing. What looks like interlevel causation really just is causation at one level combined with an interlevel constitutive relevance relation.

Problematically, though, this leaves open where the causation is happening—at the level of S's ψ-ing or at the level of X's ϕ-ing? For the mechanist, causation at all levels is possible, only it must not cross between levels. And even if we can, by some additional criterion, decide on which level to put causality, we still do not know what exactly the mechanistic interlevel constitutive relation is supposed to be. All we know is that it is non-causal in character and some kind of a part-whole relation.[1] Before proceeding to a more detailed discussion of constitutive relevance relations, let us pause here for a moment. What happens if we take the mechanists' suggestion at face value and interpret bottom-up and top-down studies in terms of Woodwardian interventions? Given that mechanists want interlevel relations to be constitutive rather than causal in nature, and that interventionism suggests a causal reading of difference making relations irrespective of mechanistic levels, we can already guess that this will be a problematic endeavor. But what exactly does it look like if we try and translate bottom-up and top-down experiments into interventionist graphs? Is there really no way of capturing interlevel manipulations in the interventionist framework? The current chapter will examine this question.

6.2 Interventionist Interpretations of Interlevel Experiments

Bottom-up and top-down experiments in the fashion of (I), (S), (A), and (D) (see p. 58) aim to *elicit* or *cause* changes (excitation or inhibition) in one domain (on one level) by means of intervening into another. Given that mechanists appeal to Woodwardian interventions in this context, it seems only plausible to try and put interlevel studies like the ones described in chapter 2.3 in interventionist terms. Eventually, we shall constrain our graphs to not include any interlevel causal relations so they accord with mechanistic theory.

In a bottom-up study, one intervenes on X's ϕ-ing to change S's ψ-ing while in a top-down study one intervenes on S's ψ-ing to change X's ϕ-ing. To detect the changes an intervention elicits, one needs to observe the phenomenon and its mechanism at time t_1 prior to the intervention and at a (slightly) later time t_2 after the intervention. I therefore conceive of X's ϕ-ing at t_1 and X's ϕ-ing at t_2 as well as S's ψ-ing t_1 and S's ψ-ing at t_2 as four variables ($\Phi_1, \Phi_2, \Psi_1, \Psi_2$) participating in the causal graphs describing cross-level experiments: for bottom-up experiments, one intervenes on X's ϕ-ing at t_1 and observes S's ψ-ing at t_2 whereas for top-down experiments one intervenes on S's ψ-ing at t_1 and observes X's ϕ-ing at t_2.

Typically, empirical studies not only measure what they intervene into and its interlevel effect but constantly assess interlevel correlates. That is, one measures both S's ψ-ing and X's ϕ-ing at t_1 and t_2. Adding in the correlations of ψ-ings with ϕ-ings, we get the causal graphs shown in figure 6.1. Dotted lines indicate correlation and curly lines indicate some form of connection to be explicated; this could be the "mechanistically mediated effect" relation.

---

1 I shall return to the issue of mechanistic constitution in chapter 7.

$\Psi_1 \quad \Psi_2$

$I \longrightarrow \Psi_1 \quad \Psi_2$

?

?

$I \longrightarrow \Phi_1 \quad \Phi_2$

$\Phi_1 \quad \Phi_2$

t

t

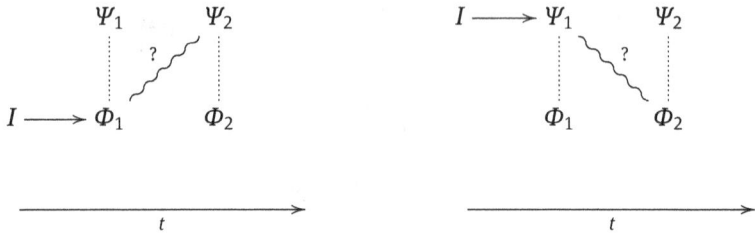

Fig. 6.1: Bottom-up (left) and top-down (right) experiments including interlevel correlates. Solid arrows indicate causal relations, dotted lines indicate correlation, curly lines indicate interlevel manipulability relations. Time is made explicit in this representation, as it is typically implicit in the way we draw interventionist graphs. The same holds for interlevel relations: Ψs are printed above Φs. Note, however, that interventionism does not come with any specific tools to handle time or levels. In fact, placement of variables should be arbitrary and not determined by such factors.

Several things should be noted at this point. First, interventionism does not actually appeal to any notion of levels; variables can be causally related quite independently of where what they represent is located in the world. I shall indicate mechanistic levels by putting variables in the same row, however. Accordingly, causal links between different rows of variables are to be understood as interlevel causes. Second, note that although the graphs in figure 6.1 somewhat resemble Kim's famous graph (see chapter 4.2.1), the interlevel relations at play are importantly different. Kim relates mental and physical properties standing in a supervenience relationship (depicted by a double tailed arrow). For now, the graphs in this chapter simply represent correlated events or processes.[2]

---

2 Nothing hinges on the exact distinction between the two here. For interventionism conceptualizes properties, events, and processes all as variables taking different values. For reasons of consistency we may demand that within one graph the variables should all represent the same kind thing. For current purposes, we may think of variables representing events (activities, or behaviors). I take processes to be chains of events. Since a phenomenon (S's ψ-ing) is defined as a mechanism's behavior, we may consider it to be an event, or, if temporally extended, a process. The same goes for entities' activities (X's ϕ-ings). Since in the current context we are discussing measures of neural and cognitive processes being taken at different points in time, it's perhaps most intuitive to think about Φs and Ψs as representing events; S's ψ-ing at t_1, for instance. However, we could also reformulate a mechanism's behavior in terms of properties. For instance, S's ψ-ing could be S's property to ψ whenever the appropriate conditions are fulfilled such that certain parts of the mechanism have certain properties. But I take this to be just a matter of wording. Since I will not be doing any heavy-duty metaphysics here, I shall not discuss this issue any further.

6.2.1 Fat-handedness

Third, the variables I am using represent a mechanisms' behavior as a whole and the activity of *individual* components in the mechanism (not the orchestrated activities of organized components). Since components should be *parts of* the mechanism, we may not place them in the same interventionist graph. This is because interventionism does not allow non-causally dependent variables to be included in the same graph. The reason for this is that non-causally dependent variables violate the independent fixability constraint (IF). The problem of including parts along with their wholes in the same graph is structurally analogous to the issue with mental causation discussed in chapter 4. If variables are related as part and whole, just as when they are related by supervenience, we cannot surgically intervene. Rather, our interventions will be *fat-handed*, viz. they will inevitably affect multiple variables at the same time. We simply cannot intervene on a whole (S's ψ-ing) while leaving all of its parts (X's ϕ-ings) untouched or vice versa. Therefore, the bottom-up and top-down experiments mechanists describe may not be "unrealizable *in principle*" (Baumgartner & Casini forthcoming, p. 2)—at least so long as we hold on to the thought that they involve proper Woodwardian interventions. For, as Baumgartner and Casini note, it seems "impossible to surgically intervene on phenomena, break constitutive dependencies, and isolate phenomenon-constituent pairs." (forthcoming, p. 10) Meanwhile, several authors have picked up on this point (see e.g. Gebharter & Baumgartner 2016; Leuridan 2012; Lindemann 2010; Mc Manus 2012; Harinen 2014; Romero 2015; Krickel 2017; Gebharter 2016; Baumgartner & Casini forthcoming) that has become known in the literature as the *fat-handedness problem*.[3] For now though, since we are currently interested in seeing what interventionism can do in the context of interlevel studies if it was possible to apply interventions here, we shall put this aside.

Assuming we can use interventionist graphs to model interlevel experiments, what really are the causal paths from Φ_1 and Ψ_1 to Ψ_2 and Φ_2, respectively, in the graphs in figure 6.1? By simply drawing out the possible paths from I to Ψ_2 and Φ_2, respectively, we arrive at the possibilities displayed in figure 6.2. In addition, combinations of these may also be at work.

Thus far, this is simply guesswork. Since according to (MM) *mutual* manipulability between the behavior of the mechanism as a whole and the activity of one of its components is sufficient for interlevel constitutive relevance, we want a single graph to picture this mutual manipulability relation. Thus, the causal graphs representing the relations between S's ψ-ings and X's ϕ-ings should—at least in

---

3 I propose a way to deal with this and other problems in chapter 11. For a recent review on different attempts to handle the fat-handedness problem see Kästner & Andersen (in press).

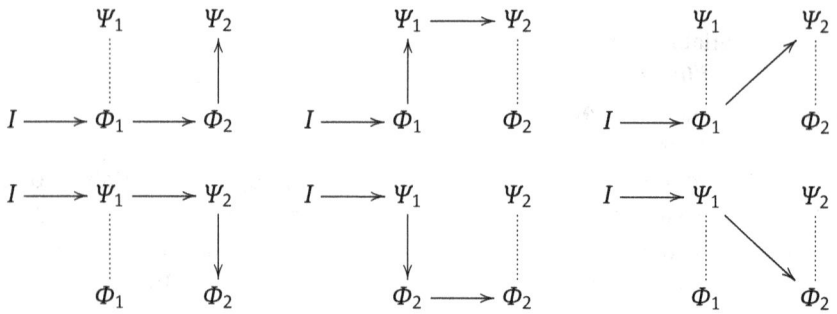

Fig. 6.2: Six possible causal graphs representing bottom-up (top row) and top-down (bottom row) experiments.

principle—be structured in such a way that they can capture both bottom-up and top-down experiments tapping into the same phenomenon and a component of its realizing mechanism. That is to say a single graph must offer both a path from Φ_1 to Ψ_2 and Ψ_1 to Φ_2. Since figure 6.2 lists the possible structures for each of these paths, we should expect the graph capturing mutual manipulability to combine one of the above graphs for top-down experiments with one of those for bottom-up experiments. Even more, if adopting the mechanistic view, we already know that the interlevel relations between Φs and Ψs cannot be causal. This renders any of the above graphs incompatible with the mechanistic view.[4] So what is the alternative?

6.2.2 Common Causes

Thus far, I only considered that the observed correlations could indicate causal links between Φ-events and Ψ-events. However, there is another possibility. According to (CP) correlated events can either be related as cause and effect or they are jointly caused by a *common cause* (see chapter 4.1).

Most straightforwardly, we could think of I as the common cause of Φ_1 and Ψ_1 where I represents both top-down and bottom-up interventions. As mentioned above, it seems odd to assume that the relation between a phenomenon and a component in its mechanism will change over time. Therefore, if we think that a

---

4 Even if there were to be causal relations between Φ_1 and Ψ_1 or Φ_2 and Ψ_2, there is no uncontroversial graph accounting for both bottom-up and top-down interventions. For brevity's sake, I shall only mention this here without argument. For the full story see p. 101.

the correlation between Φ_1 and Ψ_1 is due to a common cause, we should likewise expect the correlation between Φ_2 and Ψ_2 to be due to a common cause. This common cause could be either Φ_1 or Ψ_1 yielding the two causal graphs in figure 6.3. However, since we do not accept any graphs with interlevel causal links, we

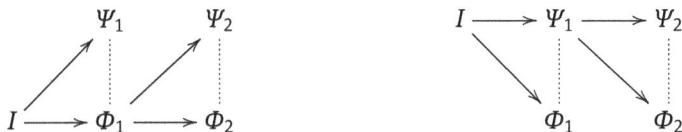

Fig. 6.3: Bottom-up (left) and top-down (right) experiments where correlating cross-level events are due to common causes.

cannot assume either Φ_1 or Ψ_1 to cause both Φ_2 and Ψ_2 for one of the the required causal links would be an interlevel link. Hence, we are left with only one possible graph to account for interlevel experiments, viz. a graph where an intervention I is a common cause of Φ_1 and Ψ_1 which in turn cause Φ_2 and Ψ_2, respectively. The corresponding causal graph is shown in figure 6.4. Note that in this graph, I is not only a common cause of Φ_1 and Ψ_1 but also a common cause of Φ_2 and Ψ_2 according to (M) and (IV). This is because I does not only cause its immediate effects but also downstream effects. I's being a common cause thus explains the correlation between Φ_1 and Ψ_1 as well as that between Φ_2 and Ψ_2.

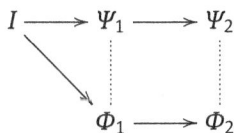

Fig. 6.4: Graph with *I* as a common cause and without any interlevel causes.

But the graph in figure 6.4 raises a curious issue: on which level shall we put I? If we put it on one level with the Ψs, the effects it has on the Φs will, once again, be interlevel effects. And if we place it on one level with the Φs, the effects on the Ψs will be interlevel effects. If we place it on a different level that neither corresponds to the level of Φs nor that of Ψs and I causes changes in both Ψs and Φs, then we have two interlevel causal links here. Since mechanists do not allow for *any* interlevel links, this renders the graph in figure 6.4 incompatible with the mechanistic account of interlevel relations as strictly non-causal.

Could we, perhaps, make a convincing case to argue that I is on both the level of Ψs and the level of Φs at once? I doubt it, for that would beg the question against the very idea of bottom-up or top-down experiments. The point of these studies is precisely to manipulate at one level and observe the effect at another.[5] However, just for the sake of the argument, let us assume there would be some kind of convincing story to this effect. But even then, we must admit that figure 6.4 is problematic. For, given that Φ_1 is always correlated with Ψ_1 (and Φ_2 with Ψ_2), and that this correlation is due to the mechanistic part-whole relation by which Ψs and Φs are related, we face a violation of (IF). We simply cannot hold the whole fixed while we fiddle with its parts or vice versa. Accordingly, once we change the value of a Ψ, this will automatically affect the value of the correlated Φ.

Another way to put this problem is to realize that the graph in figure 6.4 violates (CMC). Put briefly, the violation is that Ψ_2 and Φ_2 depend on one another (due to mechanistic constitution) although they are not not related as parent and child. This holds true at the very least as long as I cannot break the correlation between Ψ_2 and Φ_2—which it cannot because it is due to a mechanistic constitution relation.

All in all, what we can conclude from this section is that we cannot apply the interventionist framework as it stands to mechanistic top-down and bottom-up studies. This is because interventionism cannot handle graphs that include variables representing things which are related by non-causal dependence relations such as mechanistic constitution.

6.3 (M**)-(IV**)-Interventionism

I have already noted the visual resemblance of the graphs discussed here to Kim's graph. Although the cases are quite different, they share a common core: in both cases, we are trying to apply Woodwardian interventions to scenarios where there are non-causal dependence relations between variables.[6] As a result, we struggle with fat-handed interventions.

---

5 I will sketch a solution to the problem of where to locate interventions in chapter 13.

6 This is not to say mechanistic explanations face an exclusion problem in analogy to the one discussed in chapter 4.2.1. For one thing, overdetermination is not a problem for the mechanist as long as interventions screen off additional causes. For another, Craver (2007b, ch. 6) makes it quite clear that in mechanistic explanations non-fundamental features are causally and explanatorily relevant as they can be used for manipulation and control. Moreover, Kim's argument is directed at levels of realization, not levels of mechanisms which are quite different (see also chapter 3.4).

Given this resemblance between mutual manipulability and Kim's case we might try saving interventionism in the mechanistic context by applying modifications similar to those suggested to handle Kim's problem. As we still do not accept any interlevel causes (since this would be incompatible with mechanistic theory), the only possible graph to account for both bottom-up and top-down experiments is the one depicted in figure 6.4. As we have seen in chapter 4, the most promising way to handle non-causal dependence relations in interventionist graphs was to adopt (M*) and (IV*) instead of (M) and (IV). Essentially, this meant to introduce exceptions for supervening variables: they may "wiggle along" without this being interpreted causally. However, the mechanistic constitutive relation is somewhat different from supervenience. While we may think that a mechanism's behavior (S's ψ-ing) "supervenes on the organized activities of all of the components in the mechanism" (Craver 2007a, pp. 15–16), the relation of interest in interlevel studies is that between S's ψ-ing and X's ϕ-ing. The relation between these two is the mechanistic constitutive relevance relation that "holds between the phenomenon and one of the components." (Craver 2007b, p. 153, fn. 33) Thus, the interlevel relation in our graphs here is not one of supervenience.

Hence, we cannot use (M*) and (IV*) as they stand to solve our problem. What we would have to do is alter the exception clauses they contain in such a way that not only supervenience bases are allowed to "wiggle along" but also parts (if their wholes are manipulated) or wholes (if their parts are manipulated).[7] Call this (M**)-(IV**)-interventionism where (M**) and (IV**) are defined as follows:

(M**) X is a (type-level) cause for of Y with respect to a variable set V if and only if there is a possible (IV**)-defined intervention on X with respect to Y (i.e. that will change Y or the probability distribution of Y) when all other variables in Z_i in V are held fixed, except for (i) those on a causal path from X to Y, and (ii) those related to X in terms of supervenience *or a part-whole relation*.

(IV**) I is an intervention variable for X with respect to Y if and only if I meets the following conditions:
I1**. I causes X.
I2**. I acts as a switch for all the other variables that cause X. That is, certain values of I are such that when I attains those values, X ceases

---

7 To meet the mechanist's demands, we would not even have to keep the exception clauses introduced to handle cases of supervenience. However, this does not do any harm and it may be favorable for the interventionist to meet both Kim's challenge and the mechanistic one. Thus, I will keep the exception clauses introduced to handle supervenience.

to depend on the values of other variables that cause X and instead depends only on the value taken by I.

I3**. Any directed path from I to Y goes through X or through a variable Z that is related to X in terms of supervenience *or a part-whole relation*.

I4**. I is (statistically) independent of any variable Z, other than those related to X by supervenience *or a part-whole relation*, that causes Y and that is on a directed path that does not go through X.

If we also revise (IF) and (CMC) accordingly, the graph in figure 6.4 will be licensed.[8] However, there is a significant drawback to this strategy: for it to work we must already know whether there is some kind of part-whole relation between a given Ψ and a given Φ so we can correctly evaluate observed manipulability relations. But let us put this issue aside for the moment.[9]

6.4 Manipulation ... of *What*?

Let us assume that (M**)-(IV**)-interventionism allows us to handle variables representing the behavior of a mechanism as a whole and the activity of one of its component parts in one graph. Even so, the relation an (IV**)-defined intervention actually assesses will be a *causal* one. Likewise, (M**) defines what it means for some X to *cause* Y. Thus, although the graph in figure 6.4 is licensed by (M**)-(IV**)-interventionism, (M**) and (IV**) will only tell us something about the causal relations in this graph. Perhaps we can indeed use (IV**)-interventions to find that Φ_1 is causally relevant to Φ_2 and Ψ_1 is causally relevant to Ψ_2 (where we know, of course, that this apparent interlevel causal relation is really due to some kind of mechanistically mediated effects). Yet, we cannot use (M**)-(IV**)-interventionism to learn anything about the interlevel non-causal relations. It does not analyze constitutive relations but merely brackets them from its definitions. By contrast, mechanists aim to pick out precisely the *constitutive* relationships between Ψ_1 and Φ_1 and Ψ_2 and Φ_2, respectively. And they aim to do this based on the manipulability relations they observe.

The problem is even more severe than it may seem at first sight. While the graph in figure 6.4 is licensed by (M**)-(IV**)-interventionism, we cannot actually

---

8 Rejecting (CMC) and (IF) altogether does not seem plausible given that they reflect some core principles of experimental design.

9 Although we do need to know about part-whole relations to identify constitutive relevance relations according to the mechanist (cf. page 57), mechanists do not tell us how we find out about part-whole relations in the first place. I will address this problem in chapters 12 and 14.

use it to establish mutual manipulability. From (MM), we know that meeting the following three conditions is sufficient (though not necessary) for some X's ϕ-ing to be constitutively relevant to (i.e. a component of) S's ψ-ing: (i) X's ϕ-ing is a part of S's ψ-ing, (ii) there is a possible top-down intervention on S's ψ-ing with respect to X's ϕ-ing, and (iii) there is a possible bottom-up intervention on X's ϕ-ing with respect to S's ψ-ing. Now consider again the graph in figure 6.4. Granted we know that Φ_1 and Ψ_1 are related as part and whole in the mechanistic sense, what does it take to establish mutual manipulability between them? Well, we would need I to be an intervention on Φ_1 with respect to Ψ_1 and vice versa. However, we cannot get either of these: I is a common cause of both Φ_1 and Ψ_1 and therefore cannot be an intervention on one with respect to the other. This simply violates I2—irrespectively of whether we are considering (IV), (IV*), or (IV**).

As noted above, the problem at hand is a principal one. We cannot, due to the constitutive relation between S's ψ-ing (Ψ_1) and X's ϕ-ing (Φ_1), manipulate them independently. Any manipulation of the part will affect its whole an vice versa. Any intervention into either will be an intervention into the other. There are no surgical interventions, that is, but only fat-handed ones.

6.5 Upshot

Both interventionism and mechanistic explanations undoubtedly capture important and complimentary aspects of how scientist explain cognitive phenomena. As we have seen in chapter 2.3, complimentary bottom-up and top-down studies have been crucial in advancing explanations of cognitive phenomena. This is why we do want an account of scientific explanation that can handle them. While mechanistic explanations nicely capture the interlevel character of these experiments, interventionism seems suitable to capture the typically causal reading: intervening into Φ_1 affects Ψ_2 and intervening into Ψ_1 affects Φ_2. According to the mechanist, we can use Woodward-style manipulability relations to assess both intralevel causal and interlevel constitutive relationships. This makes it sound like interventionism and mechanistic explanations should be happily married.

However, my examination in this chapter has demonstrated that the two frameworks are not as easily compatible. At least not, if we are trying to use interventionism to cash out mechanistic constitutive relevance. There are principled reasons why part-whole relations cannot be assessed using Woodward-style interventions; they lie in the very definition of interventionism. Of course, this is not to say mechanistic explanations and interventionism cannot go together at all. Quite the contrary: using intralevel manipulations to determine how a mechanism's parts causally work together is, after all, uncontroversial. It is the interlevel mechanistic constitution relation and its assessment that is so troublesome for the interventionist. But this is hardly surprising given that it is a non-causal dependence relation, viz. one of those relations that Woodward does not want to be part of interventionist graphs to begin with.

The challenge for an empirically adequate and philosophically satisfying philosophy of science is to devise a theory that tells us how scientists can identify both causal and constitutive relations based on empirical manipulations. But this is not an easy task and there is quite some way to go before we will get to a solution. One of the major complications in this context is that it can be quite tricky to distinguish different types of relevance relations at all based on empirical evidence. This shall be the topic of the next chapter.

Excursus: Graphs With Interlevel Causation

Suppose, for the sake of the argument, we allowed for causation between Φs and Ψs in the causal graphs representing mutual manipulability in bottom-up and top-down studies. Consider again figure 6.2. Now, as said, we want a single graph to accommodate for both bottom-up and top-down studies.

Bottom-Up & Top-Down *Is*

Since it would be odd to assume the relation between a phenomenon and a component of its mechanism to change over time, we can rule out graphs postulating an upward causal relation between Ψs and Φs at one point in time and a downward causal relation at another (like in figure 6.5).

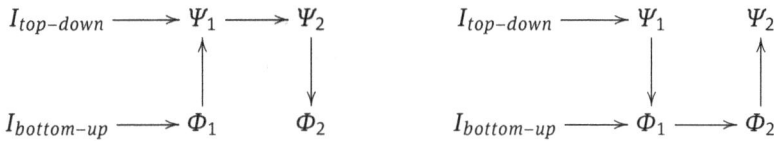

Fig. 6.5: Causal graphs combining upward and downward causation.

But we could draw causal graphs with either downward or upward connections only between Ψs and their correlating Φs (see figure 6.6). However, without further modification these cannot capture both bottom-up and top-down experiments; for either a path from Φ_1 to Ψ_2 or one from Ψ_1 to Φ_2 is missing. Even postulating an additional intralevel causal link from Ψ_1 to Ψ_2 or Φ_1 to Φ_2 would not solve this problem.

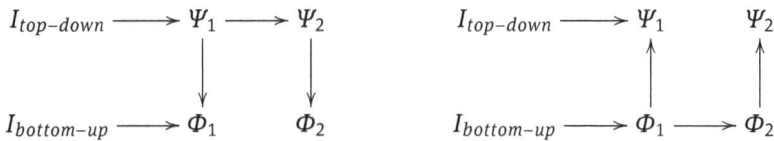

Fig. 6.6: Causal graphs with only one direction of cross-level causation.

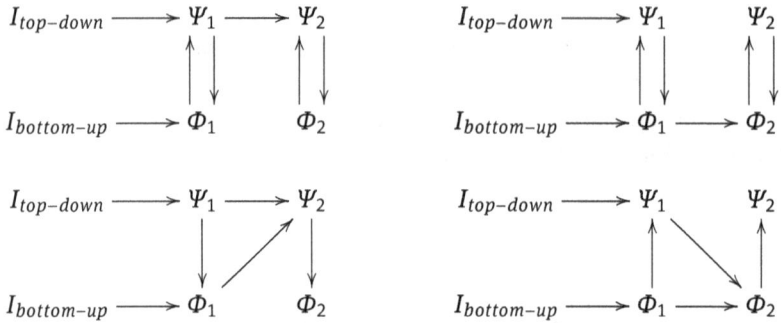

Fig. 6.7: Causal graphs with bidirectional causation between levels (top row) and direct cross-level causation (bottom row).

One way to address this problem is to introduce bidirectional connections between Φ-events and Ψ-events indicating reciprocal causation (see figure 6.7, top row). Another strategy is to introduce a direct causal link between Ψ_1 and Φ_2 or Φ_1 and Ψ_2 (see figure 6.7, bottom row). However, there are several odd things about these graphs. First, reciprocal causation is far from uncontroversial; it is ruled out by the standard conception of causation and since Woodard demands the graphs representing causal relations to be *acyclic*, the top row is certainly not an option for the interventionist. Second, the graphs in figure 6.7 postulate intralevel causal connections at one level only. However, mechanists do not seem to think of intralevel causation as restricted to specific levels. So we should probably add in intralevel causal links (see figure 6.8). Likewise we may wonder why the diagonal links should be present either from Φ_1 to Ψ_2 or from Ψ_1 to Φ_2 only. Since mechanistic levels are not systematically different and the manipulability relation is supposed to be symmetric in character, it would be more plausible to either have both of these diagonal links or get rid of them altogether. If we keep them, there will be a lot of causal overdetermination. Though not strictly forbidden according to interventionist or mechanistic views, this is undesirable. But if we do not keep the diagonal arrows, we are, again, unable to account for both interlevel manipulations within the same graph.

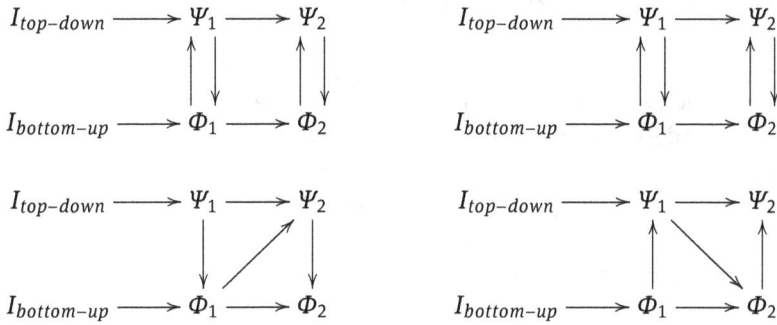

Fig. 6.8: Causal graphs from figure 6.7 with additional intralevel causal links.

I as Common Cause

If we consider possible graphs that account for the correlations between Φ_1 and Ψ_1 and Φ_2 and Ψ_2, respectively, by appealing to a common cause, we end up with two ways to capture both bottom-up and top-down studies in the same graph (see figure 6.9). As already discussed in chapter 6.2, the graph on the right hand side in figure 6.9 violates (CMC) given that we assume the correlations to stem from mechanistic constitution relations. The same reasoning applies to the graph on the left hand side. Thus, it seems, there is no plausible way of accounting for mechanistic interlevel mutual manipulability in terms of interventions.

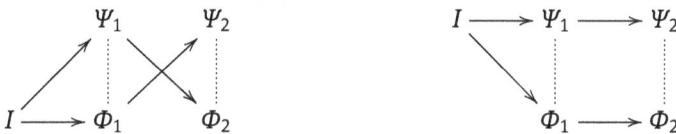

Fig. 6.9: Graph with *I* as a common cause.

7 Causation vs. Constitution

Summary
Using interventionist manipulations in mechanistic contexts is problematic; for interventionism is designed to assess causal relations and cannot handle non-causal dependence relations, such as mechanistic interlevel constitution. Causation and constitution are conceptually as well as metaphysically distinct relations that must not be conflated. Yet, typical interlevel manipulations make it quite difficult to decide which of these dependence relations is at play. Doing armchair metaphysics is of no help here; it is an empirical problem that philosophy alone cannot solve.

7.1 Two Different Dependency Relations

Thus far, I have argued that mechanistic explanations and interventionism each capture some important features of interlevel experiments in cognitive neuroscience. However, there are various complications; the perhaps most serious being that interventions in a Woodwardian fashion do not seem suited to identify—as some mechanists suggest—constitutive relevance relations (see chapter 6). This is little surprising; after all, Woodward's (2003) *interventionism* is explicitly designed as an account of *causation*, not constitution.

Constitution is, according to the mechanist, a non-causal interlevel dependence relation. As discussed in chapter 3, mechanists conceive of constitution as a *part-whole relation* between a mechanism as a whole and *one* of its component parts. The mutual manipulability criterion (MM) is intended to provide a sufficient condition for constitutive relevance. While the basic intuitions behind (MM) are quite clear, and indeed (MM) seems well-justified against the background of empirical practice, it leaves the exact metaphysics of the mechanistic constitution underdetermined.[1] No matter what the precise metaphysics, causation and (mechanistic) constitution are two conceptually quite different relations; they are even mutually exclusive. First, causes and effects must be wholly distinct while components are related to mechanisms as parts to wholes. Second, causes precede their effects but not vice versa while parts and their whole occupy the same space-time regions. Third, effects depend on their causes but not vice versa while there is a mutual dependence relation between mechanisms and their components. It is this

---

1 I will discuss (MM) in more detail in chapter 8. As we will see, there are several problems with (MM) as it stands. But for now, I will simply assume it is a plausible principle given what we see in experimental cognitive neuroscience (see chapter 2.3).

DOI 10.1515/9783110530940-007

mutual dependence relation that (MM) aims to exploit to disentangle causal from constitutive relevance relations in mechanisms.[2] Yet, we use causal manipulations in our assessment of both causal and constitutive relations.[3] Problematically, the fact that (MM) appeals to the very same Woodwardian interventions mechanists also use to identify intralevel causal relations gives the impression that these importantly different concepts get conflated.

Craver's (2007b) account of mechanistic explanations is not the only one to face this problem. Tabery (2004) identifies an analogous problem for Glennan's (2002) account of mechanistic explanations. Like Craver, Glennan adopts some crucial concepts from Woodward and applies his invariant change-relating generalizations to identify causal relations among mechanism parts as well as to identify invariance relationships between a complex system (a mechanism) and a part it contains. However, in Glennan's case the business does not seem to be quite as fishy. For him, there are nothing but causal chains (or nets) in mechanisms that we study at different levels of grain.[4] Consider a causal system that consists of a chain $A \rightarrow B \rightarrow C \rightarrow D$. Relating a change in A to how the causal chain towards D is affected, how its overall function changes, does seem just as unproblematic as (though obviously different from) relating it to how B is affected. The problem for Craver-style mechanisms, by contrast, is that they postulate a non-causal making-up relation and attempt to analyze this relation with tools designed for analyzing causal relations only (and, as chapter 6 demonstrates, are not suited for handling otherwise dependent variables).

Yet, it is tempting to analyze the interlevel manipulations we see in top-down and bottom-up studies in an interventionist fashion. After all, this is what the streamlined stories we often find in textbooks and research articles suggest. Notably, these polished stories often suggest a causal reading of interlevel manipulations. However, in most cases, the data we obtain is actually purely correlational. But the fact that we manipulate at some time t_1 and then look for an effect of

---

2 Ylikoski (2013) offers a similar distinction between causal and constitutive relations. He further emphasizes that causal and constitutive explanations not only track these different relations but also feature in different kinds of explanations relating different things. While causal explanations account for events or behaviors in terms of their causal history, constitutive explanations account for (causal) capacities, or powers, in terms of lower-level implementing structures and processes. However, for current purposes it will be sufficient to think about causal and constitutive aspects of a mechanistic explanation.

3 I will discuss this topic in more detail in chapter 11.

4 Since Glennan's notion of "levels" is quite different from the notion Craver is using, he does not run into the same problems connected with interlevel causation Craver does. See chapter 3.3.4 for more on the different conceptions of "levels". I will present an empirically guided suggestion of how to best conceive of levels in chapter 13.

our manipulation at a later time t_2 gives us a causal impression. We are "eliciting" a change in one thing through our manipulation of another, we are "making something happen"—or so it seems. The current chapter examines why interlevel manipulations look so temptingly causal and why it is so tricky to distinguish causal and constitutive dependence based on interventionist manipulations alone. This is a significant drawback of contemporary "mechanisms and interventions"-style philosophy of science. Causal and constitutive relevance are clearly different relations. It matters for an explanation whether something is a component or a cause. Mechanistic explanations and interventionism both seem to presuppose that we somehow *can* distinguish causal and constitutive relations, but as it stands they do not give us a handle on the problem of telling them apart.

7.2 Interlevel Difference Making

An ideal interlevel study applies a single manipulation at one level and assess its interlevel effect *simultaneously*.[5] This would foreclose a causal interpretation; for as causation takes time but constitution does not. Yet, to demand simultaneous manipulation and detection is often unrealistic in empirical practice. Both manipulation and measurement are subject to practical limitations and thus typically have to take place at different points in time. Similarly, scientists often assess the effects of a manipulation by comparing data obtained before and after the manipulation has been applied; or they even compare data obtained in separate experimental groups or settings.

This brings out another salient aspect of scientific practice, viz. that the effects of a given manipulation cannot always be assessed directly.[6] Instead, differences between conditions or groups are *correlated* with differences in the experimental outcome for the conditions or groups being compared. From this, we conclude that changing between groups or conditions (the difference group or condition make, respectivey) somehow "resulted in" the observed change (the measured difference) in the experimental outcome. Recall the studies described in chapter 2.3. In some neuroimaging studies (like the studies on London taxi drivers and

---

[5] While we may be able to target one (mechanistic) level specifically (see chapter 13), this does not mean the changes our manipulation induces are actually limited to that level. This is because if the mechanistic constitution relation is indeed a part-whole relation, interventions will always be *fat-handed* (see chapters 6.2.1 and 6.3).

[6] By "directly" I here mean accessed without relating it to other measured values. The question of whether or not the measurement as such is direct (as in single cell recordings) or indirect (as in most neuroimaging techniques) is a different one.

the EEG recordings during sleep) comparisons across experimental groups have been used to assess the changes in the neural domain supposedly induced by manipulations in the cognitive domain.[7] For illustration, think of MacSweeney et al. (2004) who showed participants BSL signs and gesture stimuli while their brain activations were measured. In the analysis, the researchers *subtracted* the brain activity measured during gesture processing from that measured during sign processing. Similarly, the cognitive capacities of lesion patients can be assessed pre- and post-injury, where the difference in performance is associated with the acquired brain injury. In both cases, comparisons are drawn *within subjects*, viz. differences found in the same group of subjects between the different conditions are assessed. Likewise, we can compare *between subjects* by comparing groups: in a drug trial we associate the difference in experimental conditions (e.g. treatment vs. placebo) with the differences in recovery rates between groups.[8]

In both cases, researchers rely on a so-called *subtraction logic*.[9] The experiment is designed in such a way that participants undergo (at least) one experimental and one control condition, respectively. The control condition (ideally) resembles the experimental condition in all its features except the very cognitive process (or aspect of a cognitive process) under investigation. That is, the stimuli should closely resemble one another across different tasks, the attentional resources required should be similar, and so on. This being the case, the experimenter can compare the cortical activations measured during each task and contrast (e.g. subtract) them during data analysis. The remaining cortical activation patterns are then attributed to the difference in conditions, viz. the cognitive process in question. In other words, the intervention on the higher-level (i.e. changing between control and experimental conditions) serves to induce the cognitive process being investigated while the resulting changes at the lower-level (i.e. changes in cortical activation patterns) are measured. Note that Woodward-style interventions mirror some crucial aspects of this reasoning. For instance, it is only the relevant aspect that should be manipulated while everything else is being held fixed. Yet, the subtraction method is loaded with heavy assumptions about what is "added" to a cognitive process when changing from one task to another and faces non-trivial

---

7 I will discuss the manipulative character of imaging studies in more detail in chapters 12 and 14. For now, this simplistic bottom-up and top-down manipulation picture shall suffice, even if it is not strictly empirically realistic.

8 Moreover, there are also mixed within-and-between-subject designs as well as studies using multiple interventions; see e.g. Kumari et al. (2009) or chapter 2.3.2.3. I will discuss such designs in chapter 14.

9 This *subtraction method* gets used with a wide variety of methodologies. It is not restricted to neuroimaging techniques.

methodological challenges. But for current purposes it shall suffice to see how systematic comparisons both across and within subject groups feature in interlevel experiments. This kind of reasoning is precisely the reasoning that lead to the interventionist graphs we discussed in chapter 6 (see figure 6.1). And it is this way of setting up interlevel experiments that so strongly invites a causal reading of top-down and bottom-up manipulations (irrespectively of the complication this raises for the interventionist). The reason it is so tempting to give these experiments a causal reading is that experimental practice is often quite unlike simultaneous manipulation and detection. Rather, we first measure something (a baseline condition or control group), then manipulate (set a task, run the experimental group), and then assess the difference our manipulation made.

But it is not only this temporal "skew" in the experimental reality compared to the mechanistic picture that complicates the distinction between causation and mechanistic constitution. Similarly, an ideal case of *mutual* manipulation would be one where both top-down and bottom-up manipulations can be carried out simultaneously (or at least in the same experimental setting); for this, too, would also foreclose a causal interpretation of the observed dependence (recall the conceptual distinction discussed in section 7.1). But again, this is not typically available in empirical practice. In fact, it is often difficult (if not impossible) to even use the same experimental group or method for manipulations in both directions. For one thing, different methodologies and different measurement tools may have to be applied. For another, relevant components may have to be manipulated in such a way that the converse manipulation is unavailable. Consider a lesion study. How should we elicit the cognitive process the affected area is a component for in the lesioned patient? And this is not even to speak of translational approaches that must often be used to compensate, for instance, for experiments unavailable in humans for ethical reasons.[10] Just think about the studies described in chapter 2.3 again. Scientists often use observations in clinical patients to get a first idea on what part of the brain might be relevant for what cognitive task and then they go ahead and study healthy participants, animal models, etc. (recall, for instance, the recognition memory studies in rats). They do combine evidence from complementary bottom-up and top-down studies to conclude that some brain area is actually relevant for a given cognitive process. This seems to be just what (MM) suggests. However, the experiments scientists are using to arrive at this conclusion are typically *separate* interlevel studies for bottom-up and top-down manipulations, respectively. These studies are separate in so far as they are often carried out in different experimental settings using different tools and different, though similar,

---

10 I will say more about the limits of interventionist manipulations in chapters 9 and 12.

experimental organisms. Thus, it is questionable if we can actually think of such complementary studies as involving *the same* whole and *the same* components. Besides, these studies not just employ separate *unidirectional* manipulations but, as I outlined above, they are also usually somewhat temporally extended thus inviting a causal reading for each of the applied interlevel manipulations.

However, mechanists are very explicit that there are no interlevel causal relations.[11] Neither a mechanism causes its phenomenon nor vice versa. If a constitutive mechanism is said to *produce* or *bring about* an explanandum phenomenon it does so in virtue of being constituted by its components, viz. the organized activities and entities that together build the mechanism (see e.g. Craver 2007b, p. 152). Likewise, if a change in a mechanism's behavior produces a change in a component part, then this does not actually indicate causal relation between the two. Rather, it happens just in virtue of the mechanism's being constituted by its component parts.[12] But how are we to tell such cases apart from genuine causation?

Looking at scientific practice we learn that mutual manipulation is often only *inferred* from complementary but individually *asymmetric* dependencies observed in separate experiments. This evidence alone will not be sufficient to accurately distinguish causal from constitutive relations; for it looks causal no matter what the underlying dependence relation actually is. But given that (mechanistic) constitution is supposed to be an interlevel part-whole relation, while causation is a purely intralevel relation, there are two straightforward strategies mechanists can use to disentangle the two. First, we can infer a constitutive relation when we know there is a part-whole relation. Second, we can infer an interlevel constitutive relation when we know that we actually apply intervention and detection techniques at different levels, for interlevel dependence relations cannot be causal. Either way, it seems, that distinguishing between causal and constitutive relevance requires somewhat substantial prior information about mechanisms and their (potential) components, including which entities are at the same or different levels. But neither interventionism nor mechanistic explanations tell us how to identify (potential) mechanisms and their parts to begin with. Luckily though, as we will see in chapters 12 and 14, scientists do have their ways of delineating phenomena and uncovering potential mechanisms and their components based

---

11 Note that mechanists' supposition of causation being an intralevel affair and interlevel relations being non-causal in nature is not uncontroversial. Recently, Eronen (2013) has argued that this conception is unconvincing and even incoherent. Similarly, Krickel (2017) has suggested that interlevel relations are best conceptualized as a kind of causal relation.

12 Kistler (2007) distinguishes between mechanism-to-phenomenon and phenomenon-to-mechanism relations as "constitution" and "constraint", respectively. But he too leaves the exact meaning of these terms and the metaphysics he envisons unspecified.

on non-interventionist methodologies. And they can use these to disambiguate between different dependency relations.

For now the lesson we learn is that by interventionist manipulation alone it is hardly possible to disambiguate between causal and constitutive dependence relations. Before we examine how scientists cope with this problem, however, let us try and see if perhaps understanding the precise metaphysics of the mechanistic constitution relation will be illuminating in this context. This shall be the topic of the next section.

7.3 Interlevel Mechanistic Constitution

Given the causal reading of interlevel experiments, one might think that mechanists will eventually be committed to interlevel causes after all. But as discussed in chapter 6.1, mechanists strictly deny this. Instead they argue that "interlevel causes (top-down and bottom-up) can be understood, without remainder, as appeals to mechanistically mediated effects" (Craver & Bechtel 2007, p. 547) which themselves "are hybrids of constitutive and causal relations" (Craver & Bechtel 2007, p. 562) whereof causal relations are intralevel only. But this does nothing to illuminate the constitution relation—it remains fuzzy as ever. We only know that constitution is some kind of part-whole relation and that mutual manipulability is a sufficient criterion for constitutive relevance. While the constitutive dependence relation itself is asymmetrical, we can assess it using a symmetrical manipulability relation (or at least symmetrical manipulability is supposed to be sufficient condition).

Over the past few years, the challenge to spell out mechanistic constitution in more detail has been recognized. In response, a debate about the metaphysics of this interlevel relation has developed. While this is not the place to delve into much detail on it, it is important to recognize that metaphysics is not the place to look if we aim to tease apart causal and constitutive relevance based on empirical evidence. Briefly sketching a few options of how to spell out mechanistic constitution will make it quite clear that none of the contestants solves the problem of how to empirically assess constitutive relations and distinguish them from causal ones. But this is little surprising. After all, the principle issue of constitution being an interlevel non-causal dependence remains, no matter how it is spelled out exactly.

One suggestion of how to spell out mechanistic constitution is Jens Harbecke's (2010) *minimal type relevance theory*; this account is specifically tailored to mechanistic explanations. Harbecke conceptualizes mechanisms as types, the instantiation of which are events we observe.[13] Mechanistic types are defined as minimally necessary conjuncts of types where different realizations of the same mechanistic type are expressed as disjuncts of conjuncts. A Φ-type is thus said to constitute a Ψ-type if and only if that Φ-type is an element of one of the disjuncts of possible conjuncts of types that are minimally necessary for the Ψ-type under consideration.[14] The constitution relation between processes (X's ϕ-ing and S's ψ-ing) or events (Φ_1 and Ψ_1, Φ_2 and Ψ_2) simply mirrors that between their corresponding types. If constitution is understood in this sense, then mutual manipulation is nothing over and above removing and adding types to the set constituting a mechanistic type by means of interventions.

A different, though related, proposal has been made by Mark Couch (2011). To spell out mechanistic constitutive relevance, Couch builds on Mackie's (1974) suggestion that causes can be understood in terms of complex INUS conditions, viz. insufficient but non-redundant parts of an unnecessary but sufficient condition for some effect to occur. He transfers this idea to mechanistic explanations and suggests that

> components of a mechanism that realize a capacity should be seen as *inus* components. They are the relevant parts of a structure that researchers should appeal to in giving the explanation. I propose to define a relevant part, then, as *an insufficient but nonredundant part of an unnecessary but sufficient* mechanism that serves as the realization of some capacity. (Couch 2011, p. 384)

Like Harbecke, Couch thus offers a regularity-based view of mechanistic constitution. Yet both accounts differ in important details. This is not the place to elaborate on this, but see Harbecke (2013, 2015) for a detailed comparison.

Irrespective of the very details of their definitions, neither of these accounts can solve our original problem. The reasons for this are principled: despite the merit of knowing more about the metaphysical details of mechanistic constitution, we still lack a way to *empirically* tell apart causal and constitutive relations based on the manipulability relations we observe. As long as there is *any* kind of composite-composed relation, we still have to deal with non-surgical, fat-handed interventions

---

13 Importantly, and in agreement with the mechanistic view, although certain types can constitute another type, no type *per se* is ontologically superior to another. For reasons of simplicity I will here stick to the example where ϕ-ing constitutes ψ-ing.

14 Although mechanists do not typically make this explicit, I take it that Harbecke meets their intuition in talking about mechanisms as types. This is also in accordance with the intuition that scientific explanations often generalize and thus are best considered type-level explanations. This line of reasoning is also taken by interventionists who argue that scientific explanations are type-level explanations (see chapter 4).

in interlevel experiments. And as long as we carry out interlevel experiments using the designs outlined above, we will typically have to deal with ambiguous evidence.

Another, perhaps more promising way to understand mechanistic constitution is Beate Krickel's (2014; 2017) *process interpretation* of mechanisms. According to Krickel, mechanisms are temporally extended processes. They have temporal parts (time slices) as well as spatial parts (their components). Just as mechanisms are temporally extended so are their components; they too have temporal parts. Krickel argues that within this picture we can get interlevel causation without that being a metaphysically fishy business. The clue is that given this "stretch" of mechanisms and their components along the temporal dimension we can individuate mechanisms and their components at different points in time and consider their temporal parts as wholly distinct objects. Thus it becomes possible to talk of a component's activity at time t_1 to be causally related to the mechanisms behavior as a whole at a later time t_2. Therefore, her argument goes, we can use (MM) and Woodwardian interlevel interventions to pick out components. Even though Krickel's picture may come closer to empirical practice than do other attempts to spell out mechanistic constitution, I still think this approach is problematic. First, interlevel causation as Krickel describes it essentially still is an artifact of how we intervene and detect. The interventions into a mechanism are, after all, still fat-handed and non-surgical (see chapter 6.2.1). Krickel exploits the temporal asymmetry in our methodology to fuel an intuition that is misguided from the start; the causal perception of interlevel experiments is—at least in most cases—an artifact of experimental design.[15] Second, Krickel does not provide a clear criterion of how thin or thick individual time slices should be. Given that phenomena as well as the activities of their components can vary considerably in length it might be difficult to come up with a good criterion for this. Third, mutual manipulability will—even on Krickel's view—still mostly be inferred from complementary but separate manipulative studies. Thus, identifying constitutive relations still relies essentially on knowing about parthood relations and levels before we even start to intervene. For it is this information that enables us to infer not only causal relations but also constitutive ones (and disambiguate between the two) when we observe experimental manipulability relations.

---

15 On a related note, it is important to realize that while some kind of a temporal delay is necessary for causation, it is not sufficient.

7.4 Upshot

When scientists try to explain cognitive phenomena in terms of processes in the brain they commonly use interlevel manipulations. The fact that we purposefully manipulate in one domain and subsequently assess the effects on another gives us a causal impression. However, this causal reading of interlevel manipulations leads us to conflate two crucially different concepts: causation and constitution. Just because two changes are correlated that does not mean they are causally related. The challenge is to tease apart different dependence relations that could potentially be responsible for the manipulability we observe in a given experimental setting.

But this is a tricky task. It is not one we can solve from the philosophical armchair by doing heavy-duty metaphysics. For what is required to differentiate between different dependence relations is knowledge that goes beyond manipulability and the precise details of making up relations: we need to know about mechanism parts (i.e. what our potential components are) as well as the location of different entities on mechanistic levels. Without this additional information, mutual manipulation as we find it in typical interlevel experiments will not be conclusive; it simply looks like causation.

Nevertheless, the distinction between causal and constitutive dependence is an important one. It matters for an explanation if some factor was a cause or is a constituent. I will present a way to handle this problem in part III. Before we can fully address the issue of identifying constitutive relations, however, a more detailed assessment of what mutual manipulation can and cannot do is in order. This takes us to the next chapter.

8 Beyond Mutual Manipulability

Summary
Manipulations are part and parcel of empirical research. The mechanistic mutual manipulability criterion (MM) reflects the important role that interlevel manipulations play in scientific practice. Yet, it faces a complication: empirical practice does not usually offer the clean mutual manipulations that (MM) builds on. Recently, two suggestions have been made as to how manipulability criteria might be adapted to better fit empirical practice. I will argue, however, that substantial problems remain for both. Finally, I will point to a further aspect of empirical manipulability that philosophers of science should not overlook, viz. manipulability not of components and their activities but also of their *organization*.

8.1 Vices & Virtues

Mutual manipulability is an important part of mechanistic explanation.[1] And indeed it seems well justified against the background of experimental practice. As we have seen in chapter 2.3, complementary top-down and bottom-up experiments are part and parcel of empirical research in cognitive neuroscience. Yet, as I have argued in chapter 6, (MM) seems problematic as a criterion for interlevel constitutive relevance as long as it is based on interventionist ideas about manipulations and difference making indicating causal relations. Further, as chapter 7 has shown, scientific practice cannot typically count on clean and tidy mutual manipulability. Instead, scientists often have to infer a *mutual* manipulability relationship from a series of asymmetrical, typically purely correlational, studies. The inference to constitutive relations thus relies not only on observed manipulability relations but draws heavily on knowledge about potential components within a mechanism. For it is only once we know something about the part-whole relations and the organization of the system that we can apply a criterion like (MM). Wiggling one thing (in an interventionist fashion) and seeing what else will wiggle along is not enough; for causation and constitution will both produce the same correlational (or difference making) patterns. Making inferences as to what the (mechanistic) levels are and in which direction the constitutive relation is supposed to hold, requires us to know what the parts of the whole are in the first place.

---

[1] Not all mechanistic views do in fact rely on mutual manipulability. But my discussion of mechanisms mostly builds on Craver's (2007b) account (see chapter 3).

DOI 10.1515/9783110530940-008

Since (MM) is intended to provide a sufficient criterion for constitutive relevance while we lack a necessary condition, the situation is even more troublesome. In cases where we observe manipulability in one direction only but lack complementing evidence, we will be even more tempted to infer a causal relationship; especially as long as we hold onto interventionism. The only way to foreclose this inference will be to supplement our manipulative experiments with some other kind of evidence, such as known parthood relations, etc. Note that all of these problems are quite independent of the precise metaphysics of the constitutive relation at play (see chapter 7.3).

Taken together, this shows that (mutual) difference making relations as we find them in scientific practice will often be insufficient to arrive at scientific (mechanistic) explanations. Much of the explanatory work is in fact done not by the manipulability relation but by the way it is interpreted based on additional evidence. Yet, given what we see in empirical research practice, some kind of (MM) seems important to understand how scientists construct interlevel explanations—even if it just guides our interpretation of observed manipulability. Besides, the mechanistic approach needs some kind of criterion to tease apart causal and constitutive relations; and (MM) is, albeit unsatisfying, the best we have to date. Thus, the natural question to ask at this point is what this other evidence is and how exactly we can supplement (MM) such that it becomes an empirically adequate criterion.

Recently, two suggestions have been made. Alexander Gebharter and Michael Baumgartner (2016) have suggested a criterion to supplement (MM) and get (at least indirect) evidence for a constitutive relation. Outside the debate on mechanistic explanations and interventions Alcino Silva, Anthony Landreth and John Bickle (2013) have proposed a system of experiment classification that takes into account not only difference making studies but also other kinds of experiments. I shall consider these options in turn.

8.2 Pragmatically Supplemented

Gebharter and Baumgartner (2016) set out from the problem I sketched in section 6.2.1. In a mechanism where there are constitutive relations between a phenomenon and its mechanism's components, (MM) can be satisfied only by fat-handed interventions. The fact that changes in a phenomenon are always correlated with changes in a mechanism's component does not necessarily have to be a result of constitutive dependence between the two, however. It could equally well be the case that the intervention in question just happens to be a common cause of both these changes for some other reason. Therefore, the correlation alone does not permit inference to a constitutive relation (as I have argued at length in chapter 7). Different causal models, including constitutive relations or not, might be postulated. But they will all give rise to same (the observed) correlations, i.e they will be "correlationally equivalent, and thus, empirically indistinguishable." (Gebharter & Baumgartner 2016, p. 20) We can conclude therefore, once again, that mutual manipulation is insufficient to establish constitutive relevance.[2]

To supplement mutual manipulability, Gebharter and Baumgartner introduce a pragmatic criterion that can be used to identify, at least indirectly, constitutive relevance. This criterion crucially builds on the asymmetry of the constitutive interlevel relation. As outlined earlier, the constitutive relevance relation is a relation between a phenomenon as a whole and any one of the components in its mechanism. The relation between a phenomenon as whole and the coordinated behavior of organized components might be characterized in terms of supervenience. While both these relations are asymmetric, the patterns of manipulability are different in each case. The constitutive relevance relation is characterized by mutual manipulability (this is why (MM) works as a sufficient condition). The supervenience relation, however, implies (at least on some standard reading of supervenience) that manipulations at the higher level are necessarily accompanied by changes at the lower level while manipulations at the lower level do not have to be accompanied by any changes at the higher level.

Now suppose we have a mechanism (S) with a number of components (X_1, X_2, ..., X_n). According to mechanistic theory, each of the components' activities (X_1's ϕ_1-ing, X_2's ϕ_2-ing, ..., X_n's ϕ_n-ing) is constitutively relevant to S's ψ-ing as whole and S's ψ-ing as a whole supervenes on the organized set of S's components and their activities. Thus, if we manipulate S's ψ-ing we will, due to the supervenience relation, necessarily induce some kind of change in X_1's ϕ_1-ing, X_2's ϕ_2-ing, ..., or X_n's ϕ_n-ing. By contrast, manipulations of X_1's ϕ_1-ing, X_2's ϕ_2-ing, ..., or X_n's ϕ_n-ing are not necessarily accompanied by a changes in S's ψ-ing.[3] Thus, if we carry out multiple intervention studies—in the sense of (IV*)-intervention studies apt to handle supervening variables—we should sooner or later find some surgical

2 Note that this is not to say the mechanists' (MM) is insufficient; for (MM) requires mutual manipulation *and* a part-whole relation between the relata. Once we know about the part-whole relation, mutual manipulation may be sufficient to infer constitutive relevance. However, the point here is that mutual manipulation alone will not be sufficient. The reason this is so problematic for contemporary "mechanisms and interventions"-style philosophy of science is that neither mechanistic explanations nor interventionism tell us how scientists find components before they start intervening into them. They may draw an adequate picture of something that is happening in the middle of the discovery process, but they are missing the groundwork.
3 Note that this is compatible with multiple realizability.

intervention on at least one X_i's ϕ_i-ing that is a component of S's ψ-ing. Note that this does not contradict (MM) for mutual manipulability is only a sufficient but not a necessary condition on mechanistic constitutive relevance. And, given the common assumption that (cognitive) phenomena can be implemented by different (neural) mechanisms—that they are *multiply realized*—it would not be plausible to require mutual manipulabilty as a necessary criterion. For, changes in X_i's ϕ_i-ing that do not affect S's ψ-ing could effectively just "switch" between different realizers of S's ψ-ing. Interventions on S's ψ-ing, by contrast, will never be surgical; for supervenience necessitates changes in S's ψ-ing to be accompanied by some change in at least one X_i's ϕ_i-ing that is constitutively relevant to S's ψ-ing. The non-existence of surgical interventions on S's ψ-ing, therefore, can be taken to indicate a constitutive relevance relation. Gebharter and Baumgartner call this the fat-handedness criterion (FH):

(FH)   The elements of a set $V = \{X_1, X_2, ..., X_n\}$ and a variable S satisfy (FH) if and only if every (IV*)-intervention on S with respect to some $X_i \in V$ is a common cause of S and X_i.

They suggest that the elements in V are constitutively relevant to S if and only if the relationship between the elements of V and S fulfills (MM) as well as (FH) (cf. Gebharter & Baumgartner 2016, p. 22).

While intuitively appealing, there are problems with this view. First, we cannot proof non-existence by induction. However, in empirical practice we are usually forced to draw inferences from a limited number of observations. But the mere fact that we do not find surgical (IV*)-defined interventions on S's ψ-ing is not enough to proof that no such intervention exists. It will thus be difficult, if not impossible, to gather the data required by Gebharter and Baumgartner's approach in practice. Second, carrying out all the different mutual manipulation studies required to manipulate S in such a way that different components are affected may be quite cumbersome. This may render (FH) practically infeasible. Besides, the range, precession, and sensitivity of available manipulations and measurements is limited. Methodological constraints may thus make it difficult to individually target or monitor specific components. Similarly, it may be tricky to determine whether or not a given intervention is actually fat-handed; for we may *accidentally* manipulate variables we intended to hold fixed. Third, (FH) may appear to fail if the variable set we consider is too small.[4] Suppose we have not yet discovered X_2 as a component of S. Since we do not look for changes in X_2, we will miss that some interventions into S affect X_2. If these interventions do not happen to affect

---

4 I am indebted to Markus Eronen for pointing this out to me.

any other known components of S, we are mistakenly lead to conclude that we found a surgical intervention on S. That is, we would be lead to think that (FH) fails and thus to think that the necessary condition for the remaining elements of V to be components of S is not satisfied. But this is counterintuitive: whether or not something can legitimately be counted as a mechanistic component should not depend on what other components we already know.

Gebharter and Baumgartner's criterion might *in principle* work well in some (ideal) cases. But given the practical limitations experimental research faces, it is probably best considered a *supplementary heuristic*. The same holds true for the *abductive alternative* Baumgarnter and Casini (forthcoming) suggest for (MM). In essence, their suggestion is that rather than to cash out constitutive relations in terms of mutual manipulability, we should use the impossibility of mutual manipulability scenarios as a guide to constitutive relations.

But even if either of these heuristics do work in practice, substantial problems surrounding (MM) remain. Most crucially perhaps, we still do not know how to identify the mechanism's parts which we test for constitutive relevance in the first place. Perhaps we can achieve this by using other kinds of experiments? In the next section, I consider a suggestion going into this direction.

8.3 A System of Experiment Classification?

In their recent book, Alcino Silva, Anthony Landreth and John Bickle (2013) have suggested a system of experiment classification they derive from common practices in cellular and molecular neuroscience. Note that this subfield of experimental neuroscience is significantly different from cognitive neuroscience. The most striking difference perhaps is that cellular and molecular neuroscience does not really present an interlevel science—at least not in the same obvious sense as does cognitive neuroscience. Yet, Silva, Landreth and Bickle do talk about a hierarchy of levels through which research progresses and they even discuss experiments that assess the effects of hippocampal lesions on memory performance, viz. precisely Craver's main example in his account of mechanisms.[5] This makes it plausible to assume that—although they do not explicitly discuss this—Silva, Landreth and Bickle's approach might be applied in a mechanistic interlevel context, too.

Put in a nutshell, Silva, Landreth and Bickle propose that there are three basic kinds of experiments: identity experiments, connection experiments, and

---

5 As mentioned in chapter 3.6, Bickle and Craver disagree on the (explanatory) status of higher-level entities. However, they both present realistic reconstructions of scientific practice.

tool development experiments. *Identity experiments* are designed to uncover new phenomena, describe their spatio-temporal components, and understand their properties. *Connection experiments* aim to test *causal hypotheses*; there are three different kinds: *positive manipulation experiments* that manipulate a putative cause in an *excitatory* fashion while the putative effect is being measured, *negative manipulation experiments* that manipulate a putative cause in an *inhibitory* fashion while the putative effect is being measured, and *non-intervention experiments* that just measure putative cause and effect without manipulating either.[6] To establish a causal connection, evidence from all three connection experiments should be *convergent* in the sense that exciting the cause should enhance the effect, inhibiting the cause should reduce the effect and cause and effect should be correlated outside experimental interventions. Finally, *tool development experiments* aim to develop and characterize new tools that can be used to carry out identity and connection experiments.

The reference to excitatory and inhibitory experiments is reminiscent of Craver's classification of interlevel studies into (I), (S), (A), and (D) (see section 3.4). And indeed, since both cite similar examples from memory research (e.g. hippocampal lesions) and both call for complementary evidence across different studies, readers might be tempted to think that Silva, Landreth and Bickle share a common agenda with mechanists like Craver. However, while mechanists aim to identify causal as well as constitutive relations, Silva, Landreth and Bickle emphasize that all connection experiments are "explicitly directed at probing and testing causal connections between phenomena." (Silva et al. 2013, p. 29) In fact, they do not discuss interlevel manipulations or constitutive relevance relations at all. Yet, as said, the examples they use would be considered interlevel by the mechanist. Maybe, Silva, Landreth and Bickle's neglect of interlevel manipulations just is an artifact of their ruthless reductionist convictions lingering in background (see chapter 3.6).

Be that as it may, there are three interesting points in their systematization of experiments. First, they mention purely observational (what they call non-intervention) experiments. Although these clearly contribute to the scientific enterprise, and do so quite independently of whether or not we are looking at things at different levels, neither mechanists nor interventionist talk about such experiments. Second, Silva, Landreth and Bickle mention experiments that are carried out to develop new tools. This, too, is vital to scientific practice but rarely discussed

---

6 Silva, Landreth and Bickle do not make explicit reference to any theory of causation. But the way in which they talk about manipulations somewhat resembles Woodwardian interventions. Therefore, I take it, they have in mind something along the interventionist lines.

in contemporary philosophy of science. The same goes for, third, noticing identity experiments. These experiments actually seem to be inherently interlevel studies; or at least that would be a plausible reading of identity experiments given that Silva, Landreth and Bickle talk about different portions of hippocampus performing different functions in this context.[7] Similarly, both purely observational experiments and tool development experiments could be easily transferred to an interlevel context. Thus, it might be worth trying to see if we can supplement interventionist manipulations with these different strategies in the context of mechanistic explanations. After all, the basic intuition that there are some experiments designed to discover phenomena and describe their properties and parts seems promising. Indeed, this would match the *dissociative reasoning* we discussed alongside interlevel experiments in chapter 2.3. We have seen that interlevel studies often crucially rely on discovered dissociations or get utilized to further dissociate phenomena. What sets apart Silva, Landreth and Bickle's identity experiments from other experiments is their *purpose*. These studies aim to delineate and characterize phenomena as well as their properties and parts. This is not necessarily to say such experiments actually employ a different methodology, however. And unfortunately, Silva, Landreth and Bickle do not spell out how such experiments work or what experimental options we have to design identity experiments.

Now what does this teach us with respect to mechanistic explanations, interventions and mutual manipulation? Silva, Landreth and Bickle laudably recognize that there is more to scientific experiments than manipulating something and measuring the effect: scientists are also interested in delineating phenomena and identifying their components, developing new tools and passively observing correlations. But although Silva et al. realize the need to discuss experiments serving these ends, they do not discuss *how* scientists achieve this in any systematic way.[8] Like proponents of interventionism and mechanistic explanations, in the end Silva, Landreth and Bickle mostly focus on manipulability and how it gets used to identify intralevel *causal* relations. But if we really want to understand how scientists construct scientific explanations, this cannot be the full story. For, as we have seen, there are gaping holes in our theory as long as we rely on interventions only.

---

7 Given the discussion of identity experiments, we might also think Silva, Landreth and Bickle have mind something like stages of a causal process rather than spatial parts when talking about components.

8 I will present a catalog of different experimental strategies and what research questions they can be used to answer in chapter 14.

8.4 It's Not All About Components

So far I have raised several complications with relying (more or less) entirely on interventionist manipulability to describe how scientists explain cognitive phenomena. As we have seen, using Woodwardian interventions becomes particularly difficult in an interlevel context when scientists aim to use (mutual) manipulability to assess constitutive relevance relations. A major problem in this context is that we need to know what potential components are to begin with; and just by manipulability alone, it seems, we do not get this information. But is this really true? Could we not simply poke around in a mechanism until we find something we can manipulate in the right sort of way and take this to indicate we found a component in the mechanism? I do not think so. For it is not only components of a mechanism that we can empirically manipulate. We can also manipulate their activities and organization.[9]

For illustration consider the study by Lambon Ralph et al. (2007) discussed in chapter 2.3.2.1. Using a computational modeling approach, Lambon Ralph et al. (2007) successfully mimicked two types of disease, both due to MTL degeneration: semantic dementia (SD) and herpes simplex virus encephalitis (HSVE). While SD patients typically have difficulties naming subordinate categories, patients with HSVE suffer from category-specific impairments (e.g. affecting animate items or food items only).[10] Lambon Ralph et al. systematically damaged an artificial neural network in two different ways: randomly changing the strength of the connections between the computational units (artificial neurons) mimicked HSVE symptoms while successive removal of connections mimicked SD symptoms.

Now what do we learn from this? First, knowing that MTL neurons are components in the mechanism for both subordinate category naming and identifying, say, food items does not help much to explain how the different patterns of deficits are produced. Identifying mechanistic components, that is, was not enough. Manipulating the components and their activities, e.g. by modulating activity in different

---

9 I do not doubt, of course, that sometimes scientists poke around in a mechanism to see if they can find component parts. My point is merely that once we manipulate something by poking around in a mechanism, we cannot necessarily conclude that we actually manipulated a component. Moreover, "poking around" is certainly not all that scientists do. In fact, they will usually employ more efficient and productive strategies and try to find components and reveal their organization in a more systematic way. I will say more on this in chapter 12.

10 Categories at the subordinate level are fairly specific ones, such as *finch* or *poodle* while basic level categories are more general, like *bird* or *dog*; superordinate categories are even more general, e.g. *animal*. These different levels of specificity are independent of the particular category of items consdiered.

nodes of the neural network, would not have been illuminating either. Rather, what made the difference between SD and HSVE was to manipulate the organizational structure in which the components are embedded in two different ways. The manipulations, that is, targeted the *organization*, not the components and their activities (although, obviously, the activities were affected indirectly).[11]

Organizational structures can be targeted purposefully—as in the example above—to gain additional evidence on how the mechanism under consideration works. However, it may also be an effect of erroneous experimental designs or just blindly intervening into a hypothetical mechanism. Even if we hold onto the view that scientists are using Woodward-style interventions, they might end up targeting not components *per se* but their organization if they do not pay close attention to (or just do not know precisely) where and how they intervene. This underlines the importance of finding potential components and considering with what interventions precisely we can target them before we start manipulating. Second, the study at hand clearly is an interlevel experiment (though a study in a model);[12] for we manipulate at the level of individual (artificial) neurons and their connections and observe a behavioral effect. Therefore, the current study illustrates that interlevel manipulative strategies can be used to illuminate how components are organized and causally work together. If we conceive of this example as an interlevel bottom-up experiment, we may expect, assuming appropriate methodology is available, that we can also carry out complementary top-down experiments in clinical patients. The studies in question will resemble lesion studies, but rather than investigating which brain areas are lesioned we would investigate in what way the lesions affect patients' brains. That is, for instance, we will examine how the connections between neurons are affected. Against this background it seems that some kind of mutual manipulability criterion analogous to (MM) may also be devised for identifying relevant organizational features within a mechanism. Call this the mutual manipulability criterion for relevant organizational aspects within a mechanism (MO):[13]

---

11 See Kuorikoski & Ylikoski (2013) for a related point. They argue that (MM) as stated by proponents of mechanistic explanations crucially disregards the role organization plays for bringing about a phenomenon.

12 I will say more on the use of models in part III. For now, it is sufficient to simply consider this case analogous to experiments in living organisms.

13 Of course, this criterion will only be useful when we spell out what precisely is meant by "organization" and "relevant organizational features". This makes for another discussion, however. For current purposes, simply think of the spatio-temporal properties of components and how they are linked and arranged.

(MO) An organizational aspect of a mechanism is relevant to a mechanism's be-
      havior if one can change the behavior of the mechanism as a whole by
      intervening to change the organizational aspect *and* one can change the or-
      ganizational aspect by intervening to change the behavior of the mechanism
      as a whole.

Note however, that on the mechanistic view, investigating etiological relations
including the organization of mechanistic components and their activities is the
domain of *intralevel* experiments. Once we adopt (MO), we cannot hold onto the
mechanistic dichotomy of interlevel studies investigating constitutive relations and
intralevel studies investigating etiological relations and organizational features in
a mechanism. A possible conclusion from this would be that we cannot necessarily
base a distinction between causal and constitutive relevance on whether or not
intervention and detection techniques have been applied at different levels alone.[14]
Or at least, it seems, there are cases in science where interlevel manipulation does
not serve to identify componential relations but organizational features of a causal
system.[15] Instead, what we must know is something about a mechanism's parts
(the candidate components) before we can assess the causal and/or constitutive
relevance relations they feature in.

   A related point is that in empirical practice manipulability is not necessarily
employed to study the immediate downstream effects of our manipulation.[16] Gene
knockout studies are a case in point: what is interesting about knocking out a
certain gene is not *only* the disruptions this causes in the developing organism but
also *how the organism develops differently* from a healthy organism in an effort to
compensate for the knockout. The same point can be made for lesion studies: what
is interesting about a lesion study is not only the immediate behavioral effects
but how the organism compensates for the lesion—both behaviorally and through
reorganization of the brain. I have already discussed behavioral compensations in
the context of assessing recognition memory (see chapter 2.3.1.5). For an illustration
of neural plasticity and reorganization of the brain we may consider a number of
studies involving altered somatosensory perception in individual body parts. For
instance, Woolsey & Wann (1976) examined the plasticity of somatosensory cortex
in mice. In mice, *barrel fields* in somatosensory cortex form topographic maps with
one-to-one correspondences to whiskers. The receptive fields for each whisker
can be identified via single unit recording, viz. intracranial electrode recordings

---

14 Another possible conclusion is that we should abandon the problematic notion of "levels" and
talk about "scale" and "composition" instead (cf. Eronen 2013).
15 See also chapter 6.1.
16 I am grateful to Bob Richardson for pointing this out to me.

of neural signals. When Woolsey and Wann selectively removed whiskers they found that the receptive fields of the remaining areas grow into those areas that used to be receptive fields of the removed whiskers. Missing innervation from the cut whiskers did not lead the corresponding barrel fields to die; rather, they reorganized. Similar experiments have been done in monkeys. For instance, Clark et al. (1988) found that surgically fusing two of the animals' digits by sewing them together will lead the cortical representations of the hand in somatosensory cortex to change. Initially, each finger was represented by an individual, clearly recognizable receptive field. After sewing two fingers together, the corresponding receptive fields merged to encompass both fingers. Similarly, Merzenich et al. (1984) report that after digit amputation the "representations of adjacent digits and palmar surfaces expanded topographically to occupy most or all of the cortical territories formerly representing the amputated digit(s)." (Merzenich et al. 1984, p. 591). And just as receptive fields can be diminished (or reallocated) by sensory deprivation, they can be enhanced by extensive sensory stimulation (e.g. Jenkins et al. 1990). Note though, that while observed cortical re-organization often mirrors the changes in the body, it does not necessarily do so. Amputees experiencing phantom limbs are an example where re-organization was not successful. Patients feel their limb as if it was still there, sometimes even feel sensations in other body parts as sensations of the phantom limb (see e.g. Ramachandran & Hirstein 1998).[17] To explain how and why amputees experience phantom limbs, it is important to understand how the system as a whole copes with the manipulation (how receptive fields reorganize after the amputation). That is, we need not only pay attention to downstream effects of a given manipulation (which would correspond to no signal transmission into and from the receptive fields representing the amputated limb), but also how the rest of the system works together before and after the manipulation.

It seems then that neither (MM) nor (MO) fully capture what scientists do with manipulations. Therefore, we must consider experimental practice in much more detail if we want to understand how scientists come up with explanations, especially explanations of cognitive phenomena in terms of neural processes in the brain. This will be my project in part III.

---

17 A similar phenomenon can be found in healthy people: *referred pain*. Referred pain is pain that does not occur at the site of the painful structure but in a different part of the body. A well-known case is a pain in the shoulder that is referred from the thyroid gland. The exact mechanisms behind this phenomenon are unknown but a plausible hypothesis is that the site of the referred pain corresponds to dermatomes during embryonic development.

8.5 Upshot

The mechanistic mutual manipulability criterion is based on interventionist ideas about difference making. I have argued that although the basic rationale behind (MM) is well-justified in light of empirical practices, it faces several problems. Most severely of all, (MM) presupposes that we know what potential components of a mechanism are. If we do not know this, (MM) does not get off the ground. But contemporary "mechanisms and interventions"-style philosophy of science does not tell us how to get this information. Mutual manipulation alone does not seem sufficient to pick out constitutive interlevel relations even if we suppose we can use Woodwardian interventions in interlevel contexts.

Silva, Landreth and Bickle's suggestion to look at different kinds of experiments in science seems to go into the right direction if we aim to fill this gap. In fact, John Bickle told me personally he had the tools ready to meet the challenges I described for "mechanisms and interventions"-style philosophy of science; I simply had to apply his system of experiment classification. However, I doubt the characterization of different experiments that Bickle and his colleagues offer is sufficient. They mostly focus on identifying causal relations and they seem to do this in an interventionist fashion. Although they mention other kinds of experiments they do not spell out precisely how these experiments work and which experimental strategies they employ. However, precisely this is needed if we want to understand how scientists construct explanations based on experimental evidence. Thus, I take it, Bickle and his colleagues will need something else to get off the ground—both for their own endeavor and to meet the challenges for contemporary philosophy of science I discussed here. I will provide at least some tools that can be used to this effect in chapter 12. Distinguishing different kinds of manipulations and examining how they can be systematically combined, I will develop a catalog of experiments in chapter 14 that goes significantly beyond the classification by Silva, Landreth and Bickle.

As it stands, contemporary philosophy of science with its expressed focus on interventions into mechanistic components captures only a very small portion of experimental research practice. Besides, it presupposes substantial prior information about potential componency relations. But how do we get this information to start with? As we will see in part III, scientists get this information by systematically employing various different experimental strategies. Philosophers of science will find ways to cope with the challenges I discussed once they carefully consider the details of experimental practice that common illustrations of mutual manipulability and interlevel experiments typically gloss over.

9 Interventionism's Short-Sightedness

Summary
Woodward's interventionism is designed to uncover causal relationships. While
identifying causal relations is an important part of scientific practice, there is more
to scientific explanation than just uncovering causal relations. For instance, we
may be interested in constitutive relations in addition to causal ones. But even if
interventionism is just intended to capture how scientists draw *causal inferences*,
Woodward's analysis remains incomplete. The exclusive focus on interventions in
this context is rather short-sighted.

9.1 On the Epistemology of Scientific Experiments

Interventionism has been celebrated among philosophers of science for incorporating some crucial ideas about controlling, screening off, and holding fixed possible
confounding factors that are familiar from experimental design (see chapter 4).
It is true that something like Woodwardian interventions are an important part
of scientific practice. As I have illustrated in chapter 2, scientists sometimes do
manipulate some factor X to see how this will affect some other factor Y. And
indeed such interventions on X with respect to Y can offer helpful insights into
causal processes. However, interventionism also faces several challenges. For one
thing, it lacks a clear account of how to disentangle background conditions from
experimental variables. For another, it is restricted to making inferences about
causal dependency relations and cannot handle other kinds of dependence. But
practicing scientists are not necessarily interested in causal relations only. As the
discussion of mechanistic explanations has made quite clear, we have good reason
to believe that at least some non-causal dependency relations, viz. constitutive
relations, will play an important role in at least some scientific explanations of
(cognitive) phenomena.

The interventionist could, of course, happily admit that interventionism is
tailored to uncovering causal relations only and we need some other theory to tell us
about finding non-causal dependency relations such as (mechanistic) constitution.
But even if we just aim to identify causal relations we must admit that interventions
alone can hardly be the full story of how we examine causal relations. There are
several reasons for this. First, it is quite clear that in order to use interventions
to test for a causal relationship we must first identify what the possible relata
are and learn how to target them with interventions. Interventionism does not
tell us how to achieve this. It starts from the assumption that we have a set of

DOI 10.1515/9783110530940-009

variables taking different values, but scientists have to come up with what these variables and their possible values are in the first place.[1] Second, there are different ways of making causal inferences in science and not all of them do in fact rely on interventions. There are, for instance, purely observational studies, studies involving comparisons, studies where scientists build models or start working from the structure of a system, etc. (see chapter 14).

Think, for instance about developmental psychologists assessing how *theory of mind* (ToM) skills develop in children, viz. how children acquire the ability to attribute mental states different from their own to other agents. The typical ToM task is presented verbally, illustrated with pictures, or acted by puppets.[2] said goes something like this: Sally and Anne are in a room. They find some chocolate in one of two boxes. Then Sally leaves the room and Anne removes the chocolate from one box and puts it in the other. Sally returns to the room. At this point children are asked where Sally will look for the chocolate. The correct answer is, of course, that Sally will look in the box where she saw the chocolate before she left the room. However, most children under the age of four will answer that Saidlly will look in the box where Anne placed the chocolate while Sally had been gone. This, the standard reasoning goes, is because they have not yet acquired a fully-fledged ToM (e.g. Sodian 2005; Poulin-Dubois et al. 2007).

When trying to explain how ToM is acquired scientists will typically monitor a number of different variables along with performance in the Sally-Anne-task. Such variables may include, say, children's age, their language skills, if they have siblings, whether they regularly engage with peers, how much time parents spent talking to their children, and so forth. All of these variables are observed rather than intervened into. Yet, careful data analysis will eventually lead scientists to conclude that certain aspects of a child's environment promote early ToM development while others do not. This illustrates that we can gain some form of causal knowledge without interventions.[3]

As another example think of studying how a neural signal (such as an action potential) is propagated from one neuron in a tissue sample to another. A straightforward way to do so is to find an anatomical connection along which a signal can travel, viz. a path through the network of neurons that connects the two neurons in question. This can be done, once again, without interventions. We

---

1 Note that this is quite independent of saying they must know what the "right" variables and values are, whatever that may be.

2 There are several different paradigms for testing ToM in children of different age. Some of these and their different possible interpretations are discussed in de Bruin & Kästner (2012).

3 The case of such merely observational studies is further discussed in chapter 14, especially section 14.5.1.

only need to study the architecture of the network and trace a path through it.[4] Similarly, we may learn about the path along which a signal is processed by simply *watching* it move from one point to another. We can also just watch (under powerful microscopes) how chemicals are released or bind to receptors and what changes that induces in a cell. All of this teaches us something about the causal processes within a system without having to resort to interventions. Or we may learn about the causal structure of a system by rebuilding or modeling it. For instance, we can learn about the basic processes leading up to rain by placing a cooking pot full of water on the stove, boiling it up, and observing drops of water fall from the lid back into the pot. We learn that the heat causes the vapor to rise and that once vapor collects at the top of the lid and cools down again it forms droplets of water which then fall back into the pot. In cases like these we do not need interventions to gain causal knowledge.

Similarly, we may employ systematic comparisons. Say, for instance, we want to know what caused a certain cognitive impairment in a stroke patient. How do we do this? A straightforward way to assess the structural damage that is likely to have induced the impairment is to do imaging. We compare the images of the patient's brain with those of healthy individuals (or, if available, brain images of the patient taken prior to the stroke). This will reveal which cortical structures have been damaged by the stroke. From this we can conclude that the damage to the affected areas is responsible for stroke-induced cognitive deficits.[5]

But the point is not only that it is *possible* to use non-interventionist methods to draw causal inferences. It is often necessary. For there are many cases in empirical research where clean Woodwardian interventions are impossible. This might be, e.g., for practical or ethical reasons. In ecology, for instance, paradigmatic interventionist studies would have to be widely extended in space and time and require controlling for environmental conditions that makes them almost impossible to carry out. Or perhaps we cannot use proper interventions because we do not know enough about what the relata are. Or we may not know how to control for certain variables (or that we should control for them at all) and thus cannot screen off their influence. Say, for instance, we are interested in the response properties of certain neurons in early visual areas. To study this, we can show animals stimuli with different low-level features such as bars of different orientations or blobs of different colors or illuminations and measure the neural responses in early visual areas; this is essentially what Hubel and Wiesel (1959) have done (see

---

4 I will discuss this example in more detail in chapter 12.

5 This is basically a case of subtraction (see section 7.2). Note that although this story may also be told in an interventionist-fashion, there are in fact no interventions. I will return to this in chapter 14, especially section 14.6.

chapter 2). However, since there are lots of top-down feedback connections from higher to lower visual areas, we cannot be sure that the responses we measure in lower-level areas are actually *just* immediate effects of our stimulation. Of course, we can isolate lower visual areas; but once we do this, can we be sure they still work the same way as if embedded into their anatomical surroundings? Dynamic interactions and feedback connections are highly abundant in biological systems and simply cutting them off may interfere with the very function or process we are trying to investigate. Similarly, we might lack the tools to carry out a surgical intervention such that we cannot manipulate a given variable independently of other variables. A case in point would be lesioning a deep cortical structure where lesioning that structure will mean damaging tissue on our way towards the target structure. And even if our intervention was surgical it might still change the functional relationships between unaffected variables or the overall organization of the system as it accommodates to the new situation (just as when we cut off feedback connections). In living organisms we typically find some kind of compensatory mechanism for important structures. An animal in an experimental trial might thus not even display the "pure" effect of the focal lesion the experimenters induced; it may immediately compensates for the damage by engaging other structures.

Similarly, where scientists cannot resort to animal studies they will often have to resort to studying natural lesion patients instead. But in these patients lesions are rarely ever pure; in fact, most patients have multiple clinical conditions so it becomes hard to disentangle the effects of these different conditions from one another. Or we might have to study a patient post-mortem because the structures we are interested in are not otherwise accessible. Or perhaps we will resort to animal models which requires a translational approach. In all of these cases, the system we actually study will be substantially different from the system we are trying to learn about, viz. the healthy living target organism. And even where we do not have to refer to patients or model organisms, studies may include comparisons between different groups (e.g. patients and control participants). Though such cases are typically interpreted in an interventionist fashion, there is in fact no Woodwardian intervention. Instead, scientists *pretend* they intervened when they compare groups in the analysis (the basic logic behind such comparisons is the same as in the stroke example above).

This is not to say Woodwardian interventions are of no use in science, of course. It is merely to point out that there are a number of limitations scientists face when they work with interventions. Also, I am not saying it is *only* Woodwardian interventions that face such problems; experiments without interventions are obviously subject to certain limitations, too. In empirical research, our tools—interventionist or not—are almost always imperfect. They do not usually work flawlessly and sometimes it takes multiple attempts to use them in the desired way.[6] Luckily, scientists can often compensate for the limitations of one methodology or a single experiment by combining it with another or comparing the results of multiple different experiments into a general picture (or mosaic).

The important point to get from all of this is that interventionism, even if just intended as a theory of how scientists draw causal inferences, is rather short-sighted. It neglects a wide range of experimental practices that are not based on interventions. Thus, although many empirical studies look like they are employing Woodwardian interventions, it seems inadequate to *exclusively* focus on those experiments where some X is wiggled to see if Y will wiggle along if we want to understand how scientists make causal inferences.

However, we may wonder, is it really the interventionist project to characterize the methodology behind causal inference? This takes us the the next section.

9.2 What's the Interventionist Project?

It is not quite so clear what the interventionist project is precisely. In spelling out his account, Woodward places a strong emphasis on methodology and actual empirical research practices. He does not do any metaphysics but discusses how we draw causal inferences from experimental manipulations. This has led many (myself included) to consider interventionism an account of the epistemology of causal explanation in science. And to be sure, Woodward himself describes his project as *methodological* in places. For instance, in a reply to Michael Strevens he says that his

> primary focus is *methodological*: how we think about, learn about, and reason with various causal notions and about their role in causal explanation, both as these occur in common sense and in various areas of science. (Woodward 2008b, p. 194, emphasis in original)

Insofar as this is correct and Woodward's intention is indeed to pursue an epistemic or methodological project, it is quite obvious that interventionism is rather short-sighted. As outlined above, it misses out on a wide variety of experimental practices that can be used to draw causal inferences. This is a major drawback for interventionism if it is intended as a methodological project. By focusing exclusively on interventions, interventionism cannot tell a plausible and complete story

---

6 It is not my objective to discuss the vices and virtues of different research methods. That would make for a book of its own (see e.g. Senior et al. 2006). This paragraph is intended merely to give a glimpse of the different complications individual experimental strategies might face.

of how scientists go about (causally) explaining (cognitive) phenomena. For it is missing out on some vital research methods that are, though perhaps manipulative, not manipulative in an interventionist sense.[7] Even adopting a weaker reading of interventionism according to which interventions do not necessarily uncover causal relations but dependence relations more generally (as I will suggest in chapter 11) does not solve this problem. For even when we adopt this reading, interventions still are designed to wiggle one thing to see what else will wiggle along. But this is, as the examples above clearly illustrate, not all that scientists do to draw causal inferences. And, by extension, the same holds for investigating dependence relations more generally.

But Woodward (2003) himself has also characterized his project as a *semantic* rather than an epistemic one.

> my project is semantic or interpretive, and is not intended as a contribution to practical problems of causal inference (Woodward 2003, p. 38).

If this is what Woodward is after interventionism aims to tell us what it means for a causal claim to be true.[8] It is supposed to provide us with truth conditions for certain causal (explanatory) statements. But even then, I take it that the analysis of causal claims in terms of interventions is too limited. I have just demonstrated that scientist appeal to different kinds of evidence when studying causal relations and drawing causal inferences. Should an appropriate semantics not respect this? Even if the interventionist project is a semantic one, interventionists may wish to match their semantics to the actual methodology leading up to (causal) explanations in science to make it an empirically realistic semantics.

However, Woodward also emphasizes that his project is not to specify "the full range of conditions under which causal claims can be inferred from statistical information" (Woodward 2003, p. 95). Rather, he says, he aims to specify how we test for and assess the existence of established or hypothesized causal connections. This, again, makes it sound more like an epistemological project, perhaps even a normative once. Or perhaps one that characterizes only a certain "gold standard" for testing causal claims that leaves room for other ways of inferring causal claims?

---

7 I will discuss such non-interventionist manipulations in more detail in chapter 12. There I will introduce the concept of *mere interactions* to collectively refer to all kinds of non-interventionist manipulations. A catalog of experiments involving different kinds of mere interactions will be developed in chapter 14.

8 Note that to see Woodward's account as an analysis of "causation" is rather problematic as he *presupposes* causation in his definition of interventions and vice versa (see chapter 4).

This would probably be the most favorable interpretation of the interventionist agenda.

Yet, even if interventionism just aims to explicate the "gold standard" for testing causal claims, we must wonder what other ways there are to test and infer causal relations if this "gold standard" is so often practically unavailable. In certain fields, scientists probably use microscopes and visualization techniques much more frequently than interventions. And purely observational experiments (like the ToM studies) are probably much less constrained than intervention experiments.[9] Similarly, building simple models (like the boiling water in the pot) to understand certain aspects of a phenomenon is often much more expedient than trying to manipulate the actual, possibly highly complex phenomenon. For instance, nobody would seriously consider trying to manipulate the earth's climate to explain the basic processes leading up to rain to a high school class. Therefore, even if interventionism is intended as a "gold standard" only, we might want to challenge this status. And even if we do not challenge it, we still have to wonder what other ways of accessing causal relations there are if we aim to really understand how scientists come up with causal explanations. And, of course, how other dependence relations can be uncovered and assessed if we are interested in not just causal explanations.

9.3 Upshot

I have argued that Woodward's focus on interventions is rather short-sighted. If we want to understand how scientists come up with and reason about scientific explanations—causal or otherwise—we need to understand the epistemology of scientific experiments. Woodward's interventionism is a plausible way to start because interventionism is indeed tied closely to at least some experimental practices: it does reflect established standards of experimental designs in the definition of interventions.

Yet, Woodward's analysis of causality is not sufficient if we are after an epistemology of causal inference. The crucial point is that while we may be able to demonstrate our causal knowledge by showing how we can manipulate some factor Y through interventions into some other factor X, we do not necessarily *gain* our causal knowledge in this way. We may test hypotheses about causal relations

---

9 I will say more on the different modes of scientific investigation in chapter 12. In chapter 14, I will develop a catalog of experiments classifying different studies with respect to which manipulative and observational techniques they employ.

using interventions in many—though certainly not all—cases but that does not mean we use interventions to come up with these causal hypotheses in the first place. Cases from medicine are illustrative. Finding the cause of a pain usually involves a lot of examination of different structures. Once an abnormality (say, a bone injury) is found, we can hypothesize that the patient's pain is due to this abnormality. Depending on what structures are damaged and in what way, the doctors can treat the patient, viz. *intervene*. If the treatment is successful the pain should go away. If it does not, then either the intervention was unsuccessful or the cause of the pain was not identified correctly. So either the doctors have to find another way to intervene into the putative cause or look for another possible cause. The latter, again, will likely involve examining structural aspects (e.g. taking an x-ray or MRI scan).[10]

The examples I sketched in this chapter demonstrate that understanding the epistemology of scientific experiments is much more complex than thinking about interventions into some factor X with respect to some factor Y. Even if we are interested just in how scientists draw causal inferences there is much more to be considered. In fact, Woodward's notion of interventions seems much too restrictive to cover the various experimental manipulations scientists employ; for many manipulations simply cannot meet the demands of his definition. In this context, thinking beyond interventions will be the key to a better understanding of how scientists explain (cognitive) phenomena.[11] And this will probably be true not only for causal but also for mechanistic (constitutive) explanations. Like semantics needs to be supplemented with pragmatics to truly understand the meaning of linguistic expressions, and Bayesianism needs Bayesian statistics to get off the ground, we need to supplement interventionism with other modes of scientific investigation if we want to gain a better understanding of how scientists explain (cognitive) phenomena.

A first step towards an empirically adequate epistemology of experimental research (including but not limited to an epistemology of causal inference) will be to recognize different kinds of manipulations. I will address this issue in chapter

---

10 It is true that in some cases diagnosis and treatment are not clearly separated. Also, medicine involves a lot of trial and error and sometimes interventions are employed to try to disambiguate different possible causes. For instance, a patient can be administered a neuropathic drug to find out if the cause of her pain is nerve damage that could not be otherwise assessed. But my point here is not to say interventions cannot get used in such scenarios. Rather, I want to emphasize that there are other things that get also, perhaps even primarily, used.
11 Note that I urged a similar conclusion in chapter 7. I argued that to resolve some of the problems connected to assessing interlevel constitutive relations we need to think beyond interventions, too.

12. While Woodwardian interventions are clearly important in empirical research, non-interventionist manipulations are important, too. I will introduce the concept of *mere interactions* to capture these. A second step will be to develop a catalog of experiments by investigating how different manipulations get systematically employed in different kinds of experiments and what inferences they permit. This will be the topic of chapter 14.

10 Intermezzo: Well Then?

In the second act, our main characters—i.e. interventionism and mechanistic explanation—have discovered several challenges throughout their quest to devise a plausible account of how scientists explain cognitive phenomena. The complications mostly centered around uncovering non-causal interlevel dependence relations and using Woodwardian manipulations in an interlevel context.

Mechanistic explanations have laudably moved away from the conviction that scientific explanations are merely causal in character; they introduced a constitutive component. Yet, they suggest using interventions to uncover both causal and constitutive relations. This is problematic because interventionism is designed to cash out causal relations only and explicitly requires that analyzed variables must not non-causally depend on one another. However, mechanisms do non-causally depend upon their components, viz. their *constitutively relevant parts*. To complicate matters even further, observed manipulability is effectively a black box that underdetermines which kind of dependence relation underlies it. However, as I will argue in chapter 11, we can handle these issues at rather low cost. Put in a nutshell my suggestion will be that we should see interventions not as means to uncover causal relations only but dependence relations more generally. Adopting this view— which I shall call *difference making interventionism*—renders mechanistic explanations and interventionism compatible. But even so, challenges remain for contemporary "mechanisms and interventions"-style philosophy of science.

Understanding levels and constitutive interlevel relations, for instance, is somewhat problematic. For one thing, working with interlevel manipulations to identify constitutive relations hinges upon finding part-whole relations in the first place. For another, the mechanistic conception of levels is so relativized to the very mechanism we are looking at that there is no straightforward way to integrate observations from different experiments into a single hierarchical picture or, to use Craver's metaphor, mosaic. I will argue that we can meet these challenges once we look at scientific practice more closely. After all, these questions are not only of concern to philosophers but even more so to practicing scientists. Since scientists successfully explain (at least some) cognitive phenomena, I suggest, philosophers can learn from practicing scientists how to deal with such issues.

Once we do take a closer look at what is actually going on in science, we soon come to realize that interventions are not as dominant in scientific practice as the success stories we can tell about interlevel experiments suggest. This is not to say the interventionist picture is inadequate; it certainly does capture some portion of experimental research. However, there is much more to scientific experiments

DOI 10.1515/9783110530940-010

than wiggling some factor *X* to see if some other factor *Y* will wiggle along. In fact, much of scientific practice is based on systematic observation, decomposition, and comparisons instead. Therefore, I take it, focusing on Woodwardian interventions is too short-sighted. To address this problem, I will introduce the concept of *mere interaction* in chapter 12. Mere interactions are manipulations that do not involve Woodwardian interventions. While some experiments employing such mere interactions can indeed be interpreted as if they would make use of genuine interventions (in which case I call them *pseudo-interventions*), not all of them can. I will elaborate on this when I begin to develop a catalogue of experiments in chapter 14. While I sketch ten different kinds of experiments, I do not claim this list to be complete. Neither can I pretend that my discussion covers all the aspects worth mentioning. But my aim is a modest one: I want to illuminate the very beginnings of a path along which a philosophy of experimentation can proceed to develop a full catalogue of experiments upon which an epistemology of inference to different dependence relations can be based.

Before doing so, however, I will briefly turn to the notion of "levels" in chapter 13 where I present a *perspectival view*. Essentially, my suggestion will be that "levels" are best understood as epistemic perspectives we take. This view is metaphysically non-committal, faithful to scientific practice, and recognizes that armchair philosophy is not fit to answer empirical questions.

Overall, this will demonstrate that in order to understand how scientists explain phenomena, we must study the epistemology behind different experimental strategies more closely. I shall call this branch of philosophy of science *philosophy of experimentation*. The final act takes us right into it.

Part III: **Shopping for Solutions**

"The men of experiment are like the ant; they only collect and use; the reasoners resemble spiders, who make cobwebs out of their own substance. But the bee takes a middle course; it gathers material from the flowers of the garden and the field, but transforms and digests it by a power of its own. Not unlike this is the true business of philosophy [...] from a closer and purer league between [...] the experimental and the rational [...], much may be hoped."

—Francis Bacon (1620). Novum Organum, Aphorism XCV.

11 Fixing Interventionism

Summary
Interventionism has not yet offered a convincing way of handling non-causal interlevel dependence relations. Even where we permit non-causally related variables in
the same interventionist graph, interventions are still designed to uncover causal relations only. This is not empirically adequate. I therefore suggest *difference making
interventionism* (DMI): a weaker reading of Woodward's interventionism according
to which manipulability indicates general, not necessarily causal, (explanatory)
dependence. This is plausible against the background of scientific practice and
clears the way for using interventions in interlevel experiments.

11.1 Struggling With Non-Causal Dependence

In chapters 4 and 6 we have encountered modifications of Woodward's interventionism that were designed to handle non-causally dependent variables within the
interventionist framework. The basic rationale behind (M*), (IV*), (M**) and (IV**)
was to build exception clauses into the definitions of causation and interventions
such that observed difference making is only interpreted causally if it cannot be
attributed to an already known non-causal dependence relation.

Although causal graphs including non-causal dependence relations such
as supervenience and constitution (whatever that might be precisely) are permitted between variables under (M*)-(IV*)-interventionism and (M**)-(IV**)-
interventionism, respectively, problems remain.[1] For one thing, evidence from
manipulation-induced difference making underdetermines the precise underlying
causal structure (see chapter 4.2.3). For another, we still cannot assess whether
difference making was indeed due to a causal or a non-causal dependence relation
based on manipulability. We have to know which variables are related by non-
causal dependence before we start manipulating. Only once we know this, we can
attribute the remaining difference making relations to causal relations between
the variables based on some modified version of (M) and (IV). Note that these
problems are not specific to any particular non-causal dependence relation. We
might even subject interventionism to yet another modification that incorporates
exception clauses analogous to those in (M*)-(IV*)-interventionism for all kinds

---

1 I introduce (M*)-(IV*)-interventionism in chapter 4.2.3 and (M**)-(IV**)-interventionism in
chapter 6.3. The basic idea behind both suggestions is to relativize Woodward's original definitions
of causation and intervention to dependent variable sets.

DOI 10.1515/9783110530940-011

of non-causal dependencies. Call this (M***)-(IV***)-interventionism. The core definitions would be this:

(M***) X is a (type-level) cause for of Y with respect to a variable set V if and only if there is a possible (IV***)-defined intervention on X with respect to Y (i.e. that will change Y or the probability distribution of Y) when all other variables Z_i in V are held fixed, except for (i) those on a causal path from X to Y, and (ii) those related to X in terms of *some kind of non-causal dependence relation*.

(IV***) I is an intervention variable for X with respect to Y if and only if I meets the following conditions:
I1***. I causes X.
I2***. I acts as a switch for all the other variables that cause X. That is, certain values of I are such that when I attains those values, X ceases to depend on the values of other variables that cause X and instead depends only on the value taken by I.
I3***. Any directed path from I to Y goes through X or through a variable Z that is related to X in terms of *some kind of non-causal dependence relation*.
I4***. I is (statistically) independent of any variable Z, other than those related to X by *some kind of non-causal dependence relation*, that causes Y and that is on a directed path that does not go through X.

But even (M***)-(IV***)-interventionism faces the same principled problems as (M*)-(IV*)-interventionism and (M**)-(IV**)-interventionism. For the problems arise precisely *because* non-causal dependency relations are now permitted in interventionist graphs; and without knowing where such relations obtain exactly, it becomes impossible to infer causal relations based on manipulability alone (see also Eronen2012). However, we might consider this a virtue as it matches a problem we see in empirical science: observed manipulability alone seems insufficient to disentangle different dependency relations (see chapter 7).But even so, interventionism still focuses on causal relations to do the explanatory work. Non-causal dependencies are, even on (M***)-(IV***)-interventionism, mere by-products that seem causally and explanatorily irrelevant. However, they should not be. Non-causal dependence relations can be explanatory. In fact, the whole business of finding underlying mechanisms is—at least to a certain extent—about finding constitutive relations, viz. non-causal dependence relations.

11.2 Difference Making Interventionism (DMI)

Scientific explanations are often considered primarily as *causal* explanations. In fact, the connection between causation and explanation seems to be so intimate that some philosophers even use the terms interchangeably (see Woodward 2003; Woodward & Hitchcock 2003). But there are many different kinds of explanations and many different explanatory relations in science.[2] Difference making certainly is a cornerstone of empirical research and thus of scientific explanations. But, as mechanists have laudably recognized, it is not necessarily a matter of causation, or not a matter of causation *only*. Not all manipulability, no matter how causal it looks, has necessarily a causal basis (see chapter 7.2). But how can we make interventionism work for identifying dependence relations more generally, including both causal and non-causal dependence relations?

The solution is as simple as this: modify interventionism to state that it identifies difference making relations more generally rather than exclusively causal relations. I shall call the resulting account *difference making interventionism* (DMI), or (M_{dm})-(IV_{dm})-interventionism.

(M_{dm}) X is a (type-level) *difference maker* for Y with respect to a variable set V if and only if there is a possible (IV_{dm})-defined intervention on X with respect to Y (i.e. that will change Y or the probability distribution of Y) when all other variables Z_i in V are held fixed, except for those on a *difference making path* from X to Y.

(IV_{dm}) I is an intervention variable for X with respect to Y if and only if I meets the following conditions:

     $I1_{dm}$. I causes X.
     $I2_{dm}$. I acts as a switch for all the other variables that cause X.
     $I3_{dm}$. Any *difference I makes to Y* is mediated through X.
     $I4_{dm}$. I is independent of any variable Z *making a difference to Y* and that is on a *difference making path* that does not go through X.

On this reading of interventionism difference making is not, as in Woodward's original formulation, indicative of causal relevance but of (explanatory) relevance

---

2 My discussion here focuses on causal and constitutive relations only. However, there are even more. Scientific explanations may be evolutionary, functional, computational, dynamic, etc. in character (see Mitchell 2002, 2003; Eronen 2010). How these types of explanations relate to mechanistic (i.e. causal and constitutive) ones makes for another debate (see e.g. Skipper & Millstein 2005; Glennan 2009; Woodward 2013).

more generally.[3] It may indicate a causal relation, but it may just as well indicate some kind of non-causal dependence relation, such as a constitutive relation. We can still work with directed graphs to picture difference making relations; just that on this weaker reading of interventionism we speak of *difference making paths* rather than causal paths. A difference making path in an interventionist graph may represent a constitutive relation, or a supervenience relation just as it may represent a causal one.[4] Therefore we can use (M_{dm})-(IV_{dm})-interventionism to uncover non-causal dependence relations as well as causal ones. On this view, difference making relations indicate some form of general relevance or dependence; they leave underdetermined whether it is causal or non-causal in nature. Thus we can avoid one of the major worries connected with importing interventionism into mechanistic explanations, viz. that mechanists' mutual manipulability criterion (MM) builds on Woodwardian manipulability (indicating causal relevance) to identify constitutive relations (see part II, especially chapter 8).[5] Since (M_{dm})-(IV_{dm})-interventionism does not suggest manipulability to indicate causal relations only, there is nothing wrong with using it to formulate a criterion for interlevel componency relations as well as to identify intralevel causal relations. Both cause-effect and mechanism-component relations are dependence relations.

Note that (M_{dm})-(IV_{dm})-interventionism does not completely get away without causality, though. If I is an (IV_{dm})-defined intervention on X with respect to Y the relation between I and X is still a causal one according to $I1_{dm}$.[6] The crucial

---

3 For an application of DMI to psychiatry see Kästner 2017.

4 We may want to use different arrows to distinguish between different types of relevance, but before we bother with representational features we must wonder how to distinguish different dependence relations in the first place. As we will see later (in chapter 12), this may require more than Woodwardian interventions, no matter whether they indicate causal or any other kind of relevance.

5 According to (MM), a component is *constitutively relevant* to a mechanism if and only if we can manipulate the component by intervening into the mechanism and vice versa (see p. 57).

6 It is of course possible to further modify (IV_{dm}) such that for I to be an intervention on X with respect to Y we only require that I *affects* or manipulates X in any way at all (rather than causes it). That is to say not only the dependence relation between X and Y but also that between I and X would be underdetermined. On the upside, this can accommodate for cases where I is not strictly independent of X; for instance, I could supervene on X. It would also help us avoid Woodward's circularity problem (see chapter 4.1). However, my point does not require this move. Therefore, I here stick to I's being causally related to X. (M_{dm})-(IV_{dm})-interventionism thus resembles Woodward's original account more closely than if we were to introduce further modifications. Besides, the fact that I still causes X has another advantage: it preserves our intuition that interventions are usually quite distinct from what they affect. Anyway, for current purposes nothing hinges on whether we insist the relation between I and X to be causal or permit it to be a relevance relation more generally.

point is that Y merely depends on X in one way or other. The relation between them does not have to be causal, though it might be. The general dependence that DMI picks out leaves underdetermined whether X makes a difference to Y because it is causally or non-causally relevant to Y. Given that manipulability on this reading of interventionism could potentially indicate all kinds of dependencies, there is no longer a need to exclude specific dependencies from our variable sets and graphs. Neither do we have to know where non-causal dependencies hold before we can evaluate observed manipulability relations. The downside is, obviously, that manipulability now reveals something less specific. However, this might also be considered a virtue of (M_{dm})-(IV_{dm})-interventionism as this is much more truthful to the measures empirical scientists take. As we will see in chapters 12 and 14 scientists have their ways of analyzing different dependence relations; many of which are independent of interventions. I suggest to supplement DMI with these strategies. Before we get this, however let us examine how (M_{dm})-(IV_{dm})-interventionism fares with respect to the other challenges we encountered for the different versions of interventionism (see chapter 4).

11.2.1 No Independent Manipulability

In chapter 6, I discussed at length how interventionists might interpret and handle interlevel experiments. One of the major complications was *fat-handedness*. That is, that constitutively related variables (i.e. variables representing the behavior of a mechanism as a whole, Ψs, and variables representing the activity of a component entity, Φs) cannot be manipulated independently. This does not change under (M_{dm})-(IV_{dm})-interventionism. What changes, however, is the interpretation. Since the invariant correlation between Ψs and Φs cannot be account for in terms of a causal relation on (M)-(IV)-interventionism, the only plausible interpretation is to suggest they are brought about by a common cause. But the corresponding common cause structure raised problems, too. For then an intervention I affecting a given Ψ and a given Φ would have to be such that it can affect variables at different levels. But how can that be if causation is a strictly intralevel relation?

  With (M_{dm})-(IV_{dm})-interventionism observed invariant correlations between Ψs and Φs can be accounted for simply in terms of non-causal dependence. We may think about constitution, supervenience, or identity to just name a few obvious candidates for the relation at hand. We do not have to resort to common cause structures and thus avoid all the puzzles connected with them. This is not to say there are no common causes at all, however. It is just to say we do not have to bother with causes affecting multiple levels at once. We can still make sense of fat-handed interventions (i.e. interventions that change multiple variables at once) when we think of them as interventions affecting a supervening variable the supervenience base of which will be necessarily affected in virtue of the supervening variable being manipulated. The intervention itself, though, would still only target one variable (the supervening variable) at one level. This also clears the problem of interlevel causation: we do not need it anymore to make sense of interlevel manipulability. This can be accounted for in terms of non-causal dependence relations picked out alongside causal difference making relations by (IV_{dm})-defined interventions.

11.2.2 Constitution and Causation Look Alike

As discussed in chapter 7, mutual manipulability is typically not assessed in a single study but instead inferred from individual bottom-up and top-down studies that are considered complementary. However, taken individually, each of these unidirectional (i.e. *either* bottom-up *or* top-down) studies would equally well support a causal interpretation.

Once we adopt DMI this is not surprising at all. Difference making relations can be due to causal as well as constitutive relations (or some other kind of dependence). (M_{dm})-(IV_{dm})-interventionism acknowledges the ambiguity in the evidence that manipulative studies provide. It is a virtue rather than a drawback that it does not provide us with a clear cut criterion as to how we can distinguish between causation and other forms of difference making. This is true to the experimental research practicing scientists carry out. To disentangle different kinds of dependencies, scientists can refer to additional information accessible through other methodologies. Requiring such additional evidence is not unique to (M_{dm})-(IV_{dm})-interventionism. I argued repeatedly in chapters 7 and 8 that without some way of knowing about parts within a mechanism the whole manipulability business does not get off the ground. For our manipulations have to target something to be effective in the first place. This needs more than interventions (in whatever guise); I will elaborate on such experimental strategies in chapters 12 and 14.

The fact that DMI will have to be supplemented with other experimental strategies is not a serious disadvantage over Woodward's interventionism. At least not, if the interventionist project is methodological in character (see chapter 9.2). If we want to understand how scientists explain phenomena, we will have to consider these additional strategies anyway. Besides, it is an advantage of (M_{dm})-(IV_{dm})-interventionism that it does not require us to know about non-causal dependence relations among variables before we can start to uncover difference making relations. And the fact that we can know about the relevance of certain factors for the occurrence of a given phenomenon before we know *in what way* precisely they are relevant is perfectly familiar from empirical practice.

11.2.3 Underdetermination & Distant Effects

In chapter 4.2.3 I discussed two challenges Baumgartner presented for interventionism in the context of mental causation. Even though I am not directly concerned with this topic here, I shall briefly evaluate (M_{dm})-(IV_{dm})-interventionism in light of these challenges.

The *underdetermination* problem is, in a nutshell the problem that manipulative evidence alone is insufficient to disambiguate between different causal hypotheses. Considering Kim's problem (see figure 4.3, we find that if interventions into M affect P^* this can be the case because (i) the change in M necessarily affects its supervenience base P which is causally related to P^*, or (ii) M is causally related to M^* changes in which necessarily affect its supervenience base P^*. Put slightly differently, that is to say we do not know where to locate the causal relation: between M and M^*, between P and P^*, or both?[7] Admittedly, (M_{dm})-(IV_{dm})-interventionism does not have a straightforward solution to this problem. But even so, other modifications of interventionism do not fare any better either.

The problem of *distant effects* takes issue with downstream causal effects of changes in a supervenience base (see also chapter 4.2.3). Baumgartner's (2013) analysis of (M^*)-(IV^*)-interventionism reveals that it allows for supervening variables to cause the direct causal successor of its supervenience base in a causal chain, but not any other successors further downstream.[8] Consider figure 11.1. The challenge is whether X can be considered a difference maker of Y. This is the case if and only if there is a possible (IV_{dm})-defined intervention I on X with respect to Y. Suppose I causally affects X. Since X supervenes on Z_1, Z_1 will be necessarily affected by I through X. Thus, there are two difference making paths from I through X to Y: the direct $I - X - Y$ path and the path through Z_1 and Z_2, i.e. $I - X - Z_1 - Z_2 - Y$. All the differences I makes are mediated through X and there are no other potential difference makers for X that I has to cut off. Since there are no variables in this graph that affect Y without being on a difference making path from X to Y, I is naturally independent of any such variables. Thus, I qualifies as an (IV^*)-defined intervention on X with respect to Y even though Y is not an immediate successor of the supervenience base of X, viz. Z_1. This illustrates that

---

7 For Kim, obviously, there cannot be a question as to whether P causes P^*. It does, because the physical is causally closed. But first this leaves open whether M causes M^*, and second this is not a solution the interventionist could reach based on applied interventions.

8 The problem remains even under further modifications interventionism such as (M^{**})-(IV^{**})-interventionism and (M^{***})-(IV^{***})-interventionism. This is because only immediate (synchronous) non-causal dependence relations are caught by the exception clauses introduced into (M) and (IV).

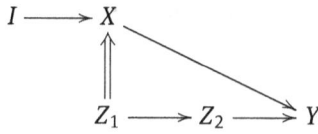

Fig. 11.1: Downstream causal effects of changes in a supervenience base are not a challenge for (M_{dm})-(IV_{dm})-interventionism. Straight arrows depict causal relations, the double tailed arrow indicates that X supervenes on Z_1; see also chapter 4.

(M_{dm})-(IV_{dm})-interventionism can successfully handle Baumgartner's problem distant effects.

11.3 Marriage Saved

Overall then (M_{dm})-(IV_{dm})-interventionism looks quite promising. It is truthful to empirical science and respects that manipulability does not necessarily imply causal relevance; it rather indicates some kind of general dependence or explanatory relevance, the precise nature of which remains underdetermined. If we adopt this reading of interventionism, a major benefit is that we can save the marriage with mechanistic explanations. For now that manipulability can indicate any kind of dependence, the most substantial problem with (MM) has been solved. Interventions are no longer designed as a tool to pick out causal relations only; they can just as well be used for identifying non-causal dependence relations. Another advantage of this non-causal reading of interventionism is that we can have feedback connections without running into major problems. On the standard reading of interventionism this used to be impossible because effects cannot influence their causes. In many biological mechanisms (such as neural networks or protein synthesis), however, feedback and recurrent connections are highly abundant (e.g. Piccinini & Bahar 2013). Once we adopt (M_{dm})-(IV_{dm})-interventionism, we can handle such mutual dependence relations. Thus, we can handle intralevel as well as interlevel recurrent structures.[9] The price is that interventionists have to give up their manifesto: there might not be causal differences without a difference in ma-

---

9 For another suggestion of how feedback connections in mechanisms might be handled while holding on to causal graphs see Gebharter & Kaiser (2014) and Clarke et al. (2014). In both cases, the basic strategies are similar: we can model feedback loops in mechanisms by building dynamic causal models "rolled out" over time. Clarke et al. (2014) use recursive Bayesian networks to do so (see also Casini et al. 2011), while Gebharter (2014) criticizes this.

nipulability relations, but *there clearly can be differences in manipulability relations without causal differences*. Other than that, we are not loosing anything substantial, except maybe the default-interpretation of difference making as causation.

But should causal relevance be the default interpretation whenever we see (unidirectional) manipulability? We have seen that this is question-begging, at least in the interlevel context. Therefore, explanatory relevance may be doing just as well in its place. After all, scientists often explain phenomena by citing relevant factors even without knowing in what way precisely these factors are relevant. Falsely postulating causal relations is certainly no better than leaving open what kind of relevance relations are at play. Rather than a tempting and problematically overused default interpretation, we need criteria that help us distinguish different kinds of relevance underlying the manipulability we observe. Requiring such additional evidence to supplement evidence from Woodward-style manipulation studies is not a vice, it is a virtue. It gives us a far more realistic picture of how scientists explain (cognitive) phenomena. They do not only use interventions, though interventions are undoubtedly a very important empirical tool. And if they use interventions, they do not necessarily do so to uncover causal connections but may be interested in other kinds of dependence relations such as constitutive relevance, organizational features and compensatory capacities of a system, or yet other information.

I already raised these issues in chapter 8 and here only reiterate the conclusions. First, interventions get used for more than testing intralevel causal claims and interlevel mutual manipulation between a mechanism and its components. Second, for the mutual manipulability criterion to get off the ground we need more than just a way to identify dependence relations. We also need a way to identify a mechanism's parts, and perhaps their organization, before we start intervening. This will give us not only the targets for our manipulations but also provide the direction for the asymmetric constitution relation we eventually want to infer from symmetric mutual manipulability.

Despite all the merits of DMI we will eventually also want to be able to reliably identify causal relations underlying observed manipulability. I believe it is unlikely that there will be single simple fool-proof strategy for this. However, there are a few heuristics we can use. Promising indications for the presence of a causal relation might include evidence that the relata are spatio-temporally distinct, that they do not normally correlate outside the laboratory, that they successively occur in a consistent order that is not permutabel, and that they can be independently manipulated where manipulating the earlier process affects the one occurring later but not vice versa, etc. (see also chapter 7). Gathering evidence like this may require using some other non-interventionist methods (see chapter 12). And obviously, it will often require induction based on a limited number of observations. But this just how it goes in science: we have only a limited number of observations, we can replicate experiments, refine our methods, develop new tools and models, but we will always have to infer our type-level explanations from a limited number of studies.

11.4 Upshot

Manipulability alone will not usually help us distinguish between, e.g., causal and constitutive relevance. However, both of these relations do some important (though different) explanatory work. Adopting (M_{dm})-(IV_{dm})-interventionism, we can assess all kinds of (explanatory) dependence relations. However, based on manipulability alone we can only capture general dependence and not which kind of dependence relation is at play precisely. Once we found a systematic dependence, we may subject it to further analyses to identify whether it is due to a causal, constitutive, supervenience, or some other kind of relation.

On this reading of interventionism importing it into mechanistic explanations is unproblematic. In fact, it is plausible that proponents of mechanistic explanations actually have something like DMI in mind when they claim interventions can be applied intra- as well as interlevel. As mentioned in chapter 3.4, Craver's notion of "production" may best be read to capture some kind of general *responsibility for* or *relevance to* the occurrence of the explanandum phenomenon. (M_{dm})-(IV_{dm})-interventionism mirrors this generality as it is suited to pick out causal as well as constitutive relations. It can thus happily marry with the mechanistic view.

The idea that difference making uncovers explanatory relevance more generally rather then causal relevance only is in line with scientific practice. In fact, scientists typically assess general dependence before they specify which kind of dependence relation is at play. And before that, they will often observe *mere correlations*. While correlations may indicate a dependence relation, the direction of this hypothetical dependence is unclear. Difference making interventions can be used to assess the direction of the dependence relation at hand. Once this is known, scientists can refer to certain inductive heuristics as well as additional experiments to disambiguate between different kinds of dependency relations that intervention studies leave underdetermined.[10] Once we consider empirical practice in more detail we will soon realize that different methods and experimental strate-

---

10 Sometimes they will have to proceed by successively eliminating different options, sometimes they will find strong evidence for the presence of a specific dependence relation. If a machine can be taken apart and its parts can be examined, for instance, we know that a given component relates to the machine as part to its whole and not as cause to effect.

gies can be put to use in this context. To recognize this, we just have to make sure to pay close attention, be aware of the practical limitations of certain experimental designs, and not let ourselves be distracted from everything else by focusing on interventions. I will get to this in chapters 12 and 14.

12 Mere Interactions

Summary

Manipulations help us understand how things work—this holds true not only for everyday life but also for scientific inquiry. Philosophers have mostly focused on Woodwardian interventions when discussing manipulations in the context of scientific explanations. But this is too short-sighted. For there is more to empirical manipulation than just Woodwardian interventions. The concept of *mere interactions* captures such non-Woodwardian manipulations. Mere interactions do not qualify as interventions in Woodward's sense; yet, they contribute crucial evidence to the explanatory enterprise in scientific practice. They thus present a substantial addition to the interventionist framework and fill a gap in contemporary "mechanisms and interventions"-style philosophy of science.

12.1 Empirical Manipulations

My discussion in the foregoing chapters has been centered around Woodward-style interventions, mostly discussing cases of interlevel manipulability (cf. chapter 2). My presentation of scientific interlevel experiments has been significantly streamlined, however. The way I presented them, these various different studies were all about wiggling something (at some given level) and assessing what else will wiggle along (at some different level) in response. This is not only how contemporary philosophy of science typically describes these experiments, it is also how the story is usually told in scientific papers. Yet, once we look at the methods cognitive neuroscientists use in more detail, we soon recognize that this picture is far too crude. Therefore, I suggest, we must examine how precisely scientific experiments work by looking at how they are designed and how the acquired data is evaluated. I shall call this strand of philosophy of science *philosophy of experimentation*.

Both mechanistic explanations and interventionism focus on a very small subset of the experimental practices scientists actually employ to construct explanations, viz. Woodward-style interventions on some factor X with respect to some other factor Y. Clearly, such interventions are a fruitful way of gathering information about dependence relations. However, they are *by far* not all that scientists can do. Besides, the observation that we can manipulate Y by intervening into X leaves the precise relation between X and Y still underdetermined. As we have seen in chapter 7 we cannot distinguish causation and constitution from the philosophical armchair. While interventionism does give us a workable account of relevance, it is not quite so clear how we are to tell apart different kinds of relevance—such as

DOI 10.1515/9783110530940-012

causal and constitutive relevance. Mechanists suggest a mutual manipulability criterion (MM) to achieve this (see chapter 3). But without knowing what potential components of a mechanism are (MM) does not get off the ground—not even if we are using certain supplementary heuristics (see chapter 8). Indeed, it will be difficult to carry out any interventions at all if we do not know precisely what to manipulate and how to target it.

Considering how practicing scientists delineate phenomena, identify underlying mechanisms and their components, and disambiguate the evidence Woodward-style manipulability offers will help us see a way to cope with these problems. Paying close attention to the experimental practices scientists employ, we will find that they do much more than wiggling X to see if Y wiggles along; they sometimes manipulate X to learn about its parts and properties quite independently of Y. The main objective of the current chapter is to introduce a distinction between these two types of manipulative strategies: classical Woodward-style interventions and manipulations that do not qualify as interventions according to the interventionist view. I shall collectively refer to such non-interventionist manipulations as *mere interactions*. Thus far, I have used the terms "intervention" and "manipulation" interchangeably. From now on "manipulation" will be considered the more general term covering both Woodwardian (or Woodward-style) interventions (i.e. manipulations that qualify as interventions according to the interventionist view) and mere interactions.[1] Recognizing the role that non-interventionist manipulations play in empirical research will be the key to a deeper understanding of how scientists explain phenomena.

In what follows I will explicate the concept of mere interactions in some detail and illustrate it by examples. For the time being I shall focus on the contrast between Woodwardian interventions and mere interactions more generally. I will discuss different kinds of mere interactions and how they get used in experimental practice in chapter 14 when I start building a catalog of experiments. For now, the lesson to learn will just be that manipulating X to affect Y is by far not the only way to gather information about dependence relations that can contribute to mechanistic (causal) scientific explanations. Empirical research (not only) in cognitive neuroscience employs a multifaceted methodology with different manipulative strategies. While mere interactions are an important part of this business, philosophers' preoccupation with Woodward-style interventions has shifted them astoundingly out of focus. Recognizing mere interactions as an important part

---

1 I do not intend to criticize the conception of events as variables being set to different values—the formalization is not at issue. In fact, I will even be using variables and causal graphs to illustrate the difference.

of the explanatory enterprise may also bear implications for our conception of causation.

12.2 Interventions vs. Mere Interactions

Put in a nutshell, mere interactions are experimental manipulations that do not qualify as interventions. The major difference between the two is how what is being manipulated is related to what is being studied. With interventions, we manipulate X to see how it affects Y. With mere interactions, we manipulate X to learn about features of X itself. Like interventions, mere interactions typically affect the system or phenomenon of interest in some way. But unlike interventions, they are not applied to see what the effects of the manipulation are going to be. Scientists know what the effects of mere interactions will be and purposefully employ them as tools. Mere interactions thus enable researchers to gather information that passive observation of nature going its own way would not reveal. Passive observations alone do not require any kind of manipulation of the thing being investigated. Mere interactions, by contrast, involve some form of manipulation, though this may be indirect. In mere interaction studies, the experimenter is doing something, be it only applying a tool or positioning an apparatus, that affects the phenomenon or system of interest. Yet, it is not the effect of that manipulation that is of interest. Figure 12.1 pictures the relation between these different modes of scientific inquiry.

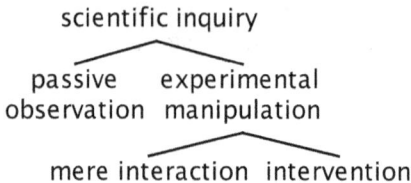

scientific inquiry

passive   experimental
observation   manipulation

mere interaction   intervention

Fig. 12.1: Modes of scientific inquiry. Passive observations contrast with manipulations which can be mere interactions or interventions.

The distinction between mere interactions and interventions is best illustrated by an example. Suppose we are given a tissue sample. For simplicity's sake let us assume that it contains at least two excitatory neurons. We already know certain basic neurophysiological facts: that neuronal signals (action potentials) are generated in the cell body once the neuron's membrane depolarizes above a certain threshold, that they are propagated along axons and received by another neurons' dendrites, and that the receiving neuron depolarizes in response to the incoming signal. Given that in our toy example there are only excitatory neurons and nothing interferes with proper signal transduction, we also know that neural signals will usually be propagated between neurons that are anatomically connected in the right way (i.e. through axon and dendrite).[2] Now suppose we want to find out whether the neurons in our sample are connected and an electrical signal will be propagated between them.

Here is one straightforward strategy to assess this: we induce a current (I) into one neuron, measure its membrane potential (M_1) and also measure the second neuron's membrane potential (M_2) where both M_1 and M_2 can take values 0 or 1 if they are below or above threshold, respectively. If I sets the value of M_1 from 0 to 1, the neuron will—given appropriate background conditions—fire. Now if we observe that once the first neuron fires the second does, too (i.e. M_2 changes from 0 to 1), we can infer, given Woodward's account, that M_1 acts as a cause of M_2 (see figure 12.2); for I is a classical Woodward-style intervention on M_1 with respect to M_2. From this, we can further infer—given our basic knowledge of neurophysiology—that there exists an anatomical connection between the two neurons. The experimental information we needed to conclude this was the change in M_2 in response to our intervention I on M_1.

$$I \longrightarrow M_1 \longrightarrow M_2$$

Fig. 12.2: Woodward-style intervention I on M_1 with respect to M_2.

Now there is another straightforward way to assess the presence of anatomical connections between the neurons in our sample: stain the tissue sample (S) e.g. by using *Golgi's method*. Bathing the sample in a potassium dichromate solution will induce a chemical process known as *black reaction* in the dendrites and soma of neurons. Thus, the otherwise opaque cell membranes can be made visible and structures belonging to a single cell can be seen in their entire length.[3] We may conceptualize that as changing the values of variables V_1 and V_2 standing for the visibility of the membranes of the first and second neuron, respectively, from

---

2 I obviously sketch a highly simplified and idealized picture here; anatomical connections are no guarantee for a signal being propagated and neither are they always necessary. But for illustrative purposes we can work with the simple picture that what fires together wires together.

3 Note that there are different staining techniques. Some can specifically target certain structures, e.g. cell bodies or myelinated fibers. As almost any scientific method, staining is an imperfect tool. For the sake of the example, however, I shall abstract away from the practical limitations and assume it works reliably and perfectly on all the cells in the sample.

values 0 (for not visible) to 1 (for visible) (see figure 12.3). Assuming our staining technique is working flawlessly and does not miss out on any axons and dendrites, this method is almost guaranteed to settle the matter of whether or not any two neurons in our tissue sample are anatomically connected, and hence, given the assumptions outlined above and that all the synapses in the sample function properly, whether an electrical signal will be propagated from one to the other.[4]

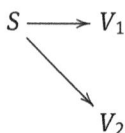

$$S \longrightarrow V_1$$
$$\searrow$$
$$V_2$$

Fig. 12.3: S is *not* a Woodward-style intervention on V_1 with respect to V_2.

One might argue that S still is an intervention on V_1 just as I is an intervention on M_1. However, the crucial difference between S and I is that while I is an intervention on M_1 *with respect to M_2*, S is *not* an intervention on V_1 with respect to V_2 (or vice versa) because S acts as a common cause of both V_1 and V_2. It thus cannot, according to Woodward's definition, be an intervention on V_1 or V_2 with respect to the other.[56]

In the first case, what we knew we would do is only change M_1; it was thus illuminating, albeit maybe hypothesized, to also see a change in M_2 resulting from I. This is what makes I an intervention on M_1 with respect to M_2: we can look for an effect in M_2 although I did not directly target it. S, on the other hand, is not an

---

4 Once again, I am assuming a highly simplified picture here. In reality, we must take into account that there are different kinds of synapses, that they can be quieted as receptors are blocked, that there are excitatory and inhibitory connections, myelinated and unmyelinated axons, that staining rarely ever catches all the neurons, etc. However, these are practical concerns that do not defeat my principle point.

5 This violates condition (iii) of Woodward's definition of an intervention: "I is an intervention variable for X with respect to Y iff (i) I causes X; (ii) I acts as a switch for all the other variables that cause X; (iii) any directed path from I to Y goes through X; (iv) I is statistically independent of any variable Z that causes Y and that is on a directed path that does not go through X." (Woodward 2003, p. 98) See also chapter 4. The problem is not *per se* that S is a fat-handed intervention (i.e. an intervention that causes many variables to change at once) but that it is precisely the relation between the co-varying variables that is at issue.

6 It is possible, of course, that S is an intervention on V_1 or V_2 with respect to some other variable representing an effect of V_1 or V_2, respectively. However, that would be quite a different scenario. I will discuss this in more detail further below.

intervention on V_1 with respect to V_2 because it affected both V_1 and V_2 directly. We knew S would affect both V_1 and V_2 in the same way. This is precisely why we did stain the cells: we wanted to be able to see them. What is illuminating about this is not the fact *that* the staining made the membranes visible; that we already knew, and it is why we employed a staining technique as tool. What is illuminating is the structure the stain reveals. Although it has been there all along, it is only after the staining that we can directly observe it and use this observation to infer (based on our basic neurophysiological knowledge) electrophysiological properties. This is why I call manipulations like S *mere interactions*. The qualification "mere" aims to emphasize the fact that we do not elicit or interfere with the very thing we are trying to observe (as it is the case when using Woodwardian interventions) but merely observe it, though perhaps using sophisticated observational tools and methods. What is illuminating about mere interactions is not that or how we manipulate a cause variable to elicit a change in an effect variable. When we use mere interactions, our aim is not to understand the connection between a manipulation and its effect; we already know what the effects will be. Instead, our aim is to observe something that is independent of the manipulation as such but that only becomes accessible once the manipulation is applied.

It is true that both I and S directly manipulate properties of the neurons in our sample. But while I elicited the neural signal propagation from the first to the second neuron by manipulating the first neuron, S targeted the sample as a whole to *merely* reveal structural information without changing anything about it. In a way, mere interactions may be conceived of as setting background conditions for our observations or as preparations for passive observations. Speaking figuratively, we might think of them as *putting something under a magnifying glass* rather than wiggling something to see what else will wiggle along. Mere interactions can be any kind of procedure or manipulation that makes accessible aspects of the system or phenomenon in question that are inaccessible to us without it. This may include organizational features of a system or information components.[7] While *mere interactions* may enable us to see things that would be otherwise hidden from our eyes, their effects are not *per se* what is interesting about them. Mere interactions do not typically affect the explanandum phenomenon. Woodward-style *interventions*, by contrast, *are* manipulations *with respect to* the explanandum phenomenon.

Importantly, mere interactions are not in any sense less valuable for constructing scientific explanations than interventions.[8] Indeed, a mere interaction like S might be considered more powerful than interventions like I. This is because it

---

7 In this sense, mere interaction is similar to passive observation without manipulation.
8 In chapter 9, I already argued that scientists gain explanatory (causal) information by different non-intervention based experiments. In chapter 14, I will say more on the different research

reveals more information at once. While *I* requires multiple attempts to reveal a whole network of anatomical connections within our sample, *S* reveals them all in one go (or at least that is what we assume in this toy example). However, the fact *that S* reveals the network is not what is exciting about *S*; for it is not the visibility of the network we are trying to explain. Mere interactions do not themselves elicit or inhibit the phenomenon of interest. Rather, they may be best considered a tool applied to prepare the system in such a way that certain features are made available for our observation. Sometimes this is sufficient to gain explanatory information; e.g. when we are looking for information about components and their organization. But sometimes mere interactions alone do not reveal explanatory information. Still, they can make important contributions to the explanatory enterprise: they may serve as tools employed to prepare for genuine Woodward-style interventions. For instance, staining a tissue sample makes it possible for us to see where the neurons are. Once we can see the neurons, we can target these individual cells to induce and measure electrical currents. Without the stain (i.e. the mere interaction), applying interventions (such as the current induction) might be difficult (if we cannot see where the cell bodies are). Thus, although the changes mere interactions bring about are not unexpected, they are important, sometimes even vital, to scientific practice—in cognitive neuroscience and beyond. But unlike interventions, which we employ to learn about their effects, we utilize mere interactions because we *know what their effects will be*. I will illustrate this peculiar relation between mere interaction and intervention more clearly when discussing the different experimental strategies in chapter 14.

Note that my distinction between interventions and mere interactions is independent of whether a manipulation involves actually touching something.[9] An intervention may consist in removing a chunk of brain by a surgeon, but it could also be a lesion induced by radiation in which case nobody would ever touch the patient's brain. Analogously, a mere interaction may involve slicing tissue and staining cells, but it could also be positioning an instrument for measurement without even touching the thing that will be measured. If we put an EEG on a participant's scalp, for instance, we do not touch her brain or neurons, but simply record their activations indirectly by recording electrical potentials on the scalp. Likewise, I am not concerned with the distinction between manipulations of actual physical systems and simulation experiments like the ones Mary Morgan (2003) talks about; whether we manipulate a model, a simulation, or the real thing is

---

questions that scientists can answer using mere interactions. In this context, I will also illustrate how different manipulative strategies get combined.

9 John Bickle and colleagues (2013) suggested a distinction between different manipulations based on whether experimenters touch the object being manipulated.

irrelevant for the distinction between mere interaction and Woodward-style intervention.[10] Also, it is not important whether we manipulate an entity, its activity, or
the organization of the whole system. Neither does my distinction depend on the
actual causal structure of the investigated system or the actual (causal) process
of manipulation. Whether something counts as a mere interaction or Woodward-
style intervention does not supervene on the causal structure it is applied to or
the causal process it takes to apply it. In fact, there is a certain relativity to our
driving research question or the explanandum phenomenon: the same manipulation on the same causal structure can be interpreted as an Woodward-style
intervention or a mere interaction *depending on what use we put it to* and which of
its effects we are interested in (see also chapter 14). This context dependency is
not my invention. It is a feature of Woodward's interventionism. It is evident from
his characterization of interventionist explanations as *contrastive*: according to
Woodward, explanations must answer *what-if-things-had-been-different* questions
(see chapter 4). Whether a manipulation is an intervention essentially depends on
whether we are *interested in the difference* (or contrast) *it produces*.

Clearly, both "intervention" and "mere interaction" are causal notions. Both
refer to types of manipulations that have some kind of effect on the investigated
phenomenon or system. The difference between them is not a causal difference.
Both I and S do, after all, exert some influence on the tissue sample (albeit on
different properties). My distinction between intervention and mere interaction
takes issue with the precise relation between what it is that is being manipulated
and what it is that is being studied. It is orthogonal to the distinction between

---

10 Scholl and Räz (2013) argue that scientists may resort to models where "direct methods of
causal inference" (as they describe interventions) are unavailable. Their idea is that if the system or
phenomenon under investigation offers insufficient epistemic access and is not suited for classical
well-controlled experiments (i.e. Woodward-style interventions), scientists can resort to models
and subject these to intervention experiments instead. At a first sight it may seem like Scholl
and Räz are articulating a distinction similar to mine: cases where scientists can intervene in
controlled experiments are separated from cases where they cannot intervene in a Woodwardian
fashion and resort to models instead. The examples Scholl and Räz discuss are drawn from ecology
and geology; as such they are naturally ill-apt for using experimental interventions (the reasons
for this are similar to what I discussed in chapter 9). While it is certainly true that scientists are
sometimes motivated to resort to models given the unavailability of experimental interventions,
this is not to say that they cannot use Woodwardian interventions in their models. Scientists
working with models can certainly examine and assess features of the phenomenon or system
being modeled without using interventions (i.e. by merely interacting with it), but they are not
limited to this. In fact, scientists often construct models precisely *because* they want to simulate
interventions unavailable in the real world. Therefore, I take it that Scholl and Räz's distinction
between controlled experiment and model-based science is independent of my distinction between
interventions and mere interactions.

interlevel and intralevel experiments. I have discussed interlevel as well as intralevel interventions in quite some detail in earlier chapters (e.g. chapter 2.3). To demonstrate the independence of the interlevel-intralevel distinction from the mere interaction-intervention distinction, we further need to establish that there are mere interactions in intralevel as well as interlevel contexts. As an example of an *intralevel mere interaction* consider inducing a tracer into a cell body. The tracer will subsequently be absorbed by connected cells. Visualization of the tracer will make visible the structures into which the tracer has been absorbed. Thus, this mere interaction reveals how cells are arranged and connected. To illustrate *interlevel mere interactions* we might think about putting a tissue sample under a microscope. Once subjected to bright light and enormously enlarged, researchers can make out individual cells and perhaps even components in the cells. That is, they can identify parts that are—in a mechanistic as well as an intuitive sense—on a lower level than the tissue sample as a whole that we merely interacted with.

All in all, it is perhaps little surprising that there is more to experimental research than intervening into X to affect Y. I have already illustrated that interventionist studies rarely stand alone in science. In fact, they often go hand in hand with conceptual analysis and dissociative reasoning (see especially chapters 2.3 and 8.3). I have also pointed out that scientists are often forced to infer Woodwardian manipulability from indirect evidence where they cannot actually test it by experiment—be it for practical, methodological, or ethical reasons (see chapters 7.2 and 9). In such cases, scientists resort to alternative experimental strategies to gain explanatory information. These are instances of *mere interactions*. Some experiments based on mere interactions may include systematic comparisons to simulate interventions. Although the manipulations at hand might actually not have been proper interventions, it is sometimes plausible to describe such experimental designs in terms of Woodward-style interventions and evaluate the experiments *as if* there had been genuine Woodward-style interventions. I shall refer to mere interactions interpreted in an interventionist fashion as as-if-interventions, or *pseudo-interventions*. However, not all mere interactions qualify as pseudo-interventions.[11] And those manipulations that cannot straightforwardly be given an interventionist interpretation are particularly interesting in light of the complications contemporary "mechanisms and interventions"-style philosophy of science faces. For they often give us the very kind of information we desperately seek to be

---

11 Pseudo-interventions are *a way of interpreting sets of mere interactions*. As such, we may consider them as a separate kind of manipulation, although they are reducible to mere interactions. I will discuss pseudo-interventions in more detail in chapter 14.

able to (i) apply interventions and (ii) disambiguate the evidence Woodwardian wiggling studies provide.[12]

12.3 Causal Pluralism?

I have argued that interventions and mere interactions both have their place in the explanatory enterprise. Now if explanatory information can be uncovered by these different kinds of manipulations, a natural question to ask is this: Do different kinds of manipulations uncover different (explanatory/relevance) relations? As I described above, mere interactions are particularly useful for delineating phenomena, identifying underlying mechanisms and their components, and finding part-whole relations. Intervention-based research, by contrast, is particularly valuable for identifying relations of manipulability. To identify what to intervene into and to disambiguate what kind of dependency relation (a causal or a constitutive one) underlies an observed manipulability relation, we can use evidence from mere interaction experiments. But scientists can also learn about causal relations by mere interactions alone. The example of inducing a tracer to visualize connections between cells is a case in point: we merely interact to collect information about not only structural features but also *causal processes* at a given level. For it is some kind of metabolic or other causal process that will take the tracer along and distributes it to connected cells.

In chapter 9, I briefly discussed the possibility of interventionism being a semantic account of causation. Now let us follow this idea for a moment. Suppose interventionism is in fact intended as a semantic or interpretive project telling us what it means for X to cause Y. Also, for the sake of the argument, ignore the interlevel business and different dependence relations playing into explanatory relations. Suppose instead, that we are just interested in intralevel causal claims and their semantics. Now recognize that even if we constrain our attention to intralevel causal relations we must acknowledge—given what we just learned about different manipulations—the role that mere interactions play to uncover them. In fact, appealing to physical processes to learn about causal relations is reminiscent of Salmon's causal-mechanical (C-M) model of explanation. As discussed earlier (in chapter 3.1), Salmon demands physical contact for causal (explanatory) relations which leads him to, e.g., look for anatomical connections using tracers to learn

---

12 Obviously, the dependence relation that Woodwardian interventions indicate only can be disambiguated once we acknowledge that interventions uncover some kind of general rather than a merely causal dependence. I argued for this in chapter 11 where I suggested that interventionism is best understood in terms of (M_{dm}) and (IV_{dm}) (see p. 141).

about how a system's parts are related. However, we have seen in chapter 4.1 that Salmon's C-M model leaves him tragically without a workable account of relevance. This is why Woodwardian interventions have been so successful in philosophy of science: they did not have to rely on such things as physical contact or the transmission of a mark. Yet, it is a curious fact that Woodward disregards exactly those cases on which Salmon builds his view; leading him (Woodward) to forget about mere interactions and physical structures over interventions. As a result, one might argue that the interventionist framework is left similarly incomplete as Salmon's C-M model, just in a different respect.

Recognizing the importance of non-interventionist manipulations in scientific practice, it may be worthwhile to rethink our attitude towards both interventionism and Salmon's C-M model. If the manifestations of causal relations are diverse so should be our assessment of them. Given the vast variety of methodologies that scientists apply in this enterprise, it is perhaps unlikely to find a single set of generally applicable criteria to identify causal relations. This might indicate that we are actually looking at many different things when we consider "causation". Perhaps "causation" is some kind of cluster concept und which different types of causal relations find shelter.[13]

In recent works, Glennan (2009; 2010b) laudably recognizes that the diagrams we draw when working with interventions and mechanisms obscure the richness of causal (or perhaps also non-causal dependence) relations involved in scientific explanations. He develops a view according to which there are two different kinds of causal relations at play in mechanisms he calls "causal relevance" and "causal productivity". Put in a nutshell, *causal productivity* relates events where events are best understood as objects doing something. Insofar as causal productivity relies on *de facto* physical connections, we may also think about it as *material causation*. It is a causal productive relation that Machamer, Darden and Craver's *productive continuity*, Glennan's earlier *interactions*, and Salmon's *mark transmission* capture.[14] By contrast, *causal relevance* is a relation between causally relevant factors such as properties of events, or background conditions and events. It is a counterfactual notion picking out what it is about a certain event that makes it cause another. Importantly, we can have independent knowledge about these

[13] There are many varieties of causal pluralism (see e.g. Hitchcock 2007; Godfrey-Smith 2012). However, this is typically not discussed in connection with "mechanisms and interventions"-style philosophy of science.
[14] Note that Glennan's notion of production is quite different from the one invoked by Craver (2007b) where the (lower-level) mechanism as a whole *produces* a (higher-level) phenomenon. This is also the second sense in which MDC use "production": as a kind of interlevel making-up relation (cf. chapter 3.3.4).

relations. For instance, we might know that Jesse's coming home causes his dogs to wag their tails. But we do not know what it is about Jesse's coming home that makes the dogs behave in this way. Are they happy he is finally back? Or are they begging for food? Do they want him to take them for a walk? Is it a reflex induced by smelling their master? In such a case, Glennan argues, we know something about causal productivity without knowing about causal relevance. Conversely, we might have information about causal relevance while lacking knowledge about causal productivity. We may know, for instance, that low serotonin levels are causally relevant to depression. So when a patient with low serotonin develops a depression we have good reasons to believe that low serotonin was relevant to her developing a depression. Yet, we do not know what particular event lead to the onset of the depression (cf. Glennan 2010b, p. 366).

According to Glennan, a full understanding of the causal basis of some phenomenon requires us to know about both causal relevance and causal productivity. Only if that is the case we can give a satisfactory scientific explanation. Though the notions of causal productivity and causal relevance are distinct, they are related: "events that produce an effect are also causally relevant to that event" (Glennan 2009, p. 328) while the converse does not hold. This becomes clear once we consider cases of causation by omission or disconnection (see chapter 3.3.2). In these cases, causally relevant events are absent; they cannot be causally productive. Nonetheless, certain other events may counterfactually depend on their absence. A case in point would be an alarm that goes off when the connection between a sender and a receiver unit fails. Normally, the sender would send a control signal to a receiver. As long as the connection is intact, the signal will get to the receiver and the alarm is suppressed by the control unit. But if the control signal does not arrive at the receiver, say because a door has been left open, the alarm will go off since the control unit will no longer suppress it. Here not-receiving the control signal is causally relevant to the alarm's going off although we could not speak of it being causally productive.

Note that Glennan's distinction between causal relevance and causal production is different from that between Woodward-style and Salmon-style causal relations. Only a subset of Glennan's causal productive relations qualifies as causal relations on Salmon's view: those that rely on actual physical connections. By contrast, both causal relevance and causal production as Glennan characterizes them qualify as causal relations on Woodward's view; for they both exemplify difference making relations. However, both Glennan's distinction as well as the contrast between Woodward-style and Salmon-style causal relations make it quite evident that explanations may require more than one unified conception of causal relations.[15] While I motivated these considerations given that we can identify causal relations using mere interactions as well as interventions, my distinction between mere interactions and interventions is quite independent of what different causal relations there are. Besides, I do not propose this distinction to analyze causal relations but to sketch an empirically adequate view of how scientists explain phenomena—quite independently of whether the explanatory dependence relations they discover are causal in nature. This brief detour into the semantics and metaphysics of causation has demonstrated that paying close attention to scientific methodology, acknowledging the pluralism of research strategies scientists employ, and thinking beyond interventions may bear significant implications for many different discussions in philosophy of science. But as said, for the purpose of this book I shall focus on scientific methodology and the epistemology of scientific experiments.

12.4 Upshot

Wiggling X to see if Y wiggles along is not all scientists do when they experimentally manipulate. They may just wiggle X to see what happens to X or what falls out of X when it is wiggled, put X under a magnifying glass, look at it from different perspectives, or cut X into pieces to examine it more closely. However, philosophers of science have focused extensively on Woodwardian wiggling interventions. One reason for this might be philosophers' focus on relevance relations rather than physical contact in light of challenges leveled against previous theories of scientific explanation such as Salmon's C-M model. Another might be enthusiasm about successful wiggling-type experiments we see in scientific practice. Yet, as I argued in part II, the presentation of empirical research in terms of neat bottom-up and top-down interventions is highly idealized. Besides, interventionist manipulability studies alone are usually insufficient to construct scientific explanations. In order to intervene into anything at all, we must know where and how to manipulate as well

---

15 This is not the place to enter into a detailed discussion but note that there are a few suggestions to this effect in the literature. Think for instance about Dretske's (1993) well-known distinction between *triggering* and *structuring* causes. While triggering causes may be viewed in analogy to Glennan's causal productivity, structuring causes are perhaps more like Glennan's causal relevance. Campagner & Galavotti (2007) propose, though on different grounds than I do, that there is room for both Salmon-style and Woodward-style causal relations in mechanistic explanations. Similarly, Ned Hall (2004) argues for the ineliminability of both a productive and a counterfactual notion of causation. Finally, Waters (2007) suggests to distinguish between *actual* and *potential* difference makers (see footnote 31 on p. 209).

as find a way to disambiguate between different relevance relations manipulability can indicate.

As discussed in chapter 8.3, Silva, Landreth and Bickle (2013) notice these needs at least to a certain degree when they suggest there are certain kinds of experiments that are designed to identify the components and properties of phenomena and develop new research tools.[16] But they fail to spell out what it is that scientists do when they do not intervene in a Woodwardian fashion and what the corresponding experiments look like.

Mere interactions are powerful tools that can fill this gap. They can supplement Silva et al.'s view as well as mechanistic accounts of scientific explanations. For mere interactions provide a way to find a mechanism's parts. Only once we know what potential components are manipulations can be put to work. Any kind of mutual manipulability criterion such as (MM), perhaps supplemented with certain heuristics like the one suggested by Gebharter & Baumgartner (2016) (see chapter 8.2), will only get off the ground provided that we have a way of finding out what to target with our interventions. Besides, mere interactions can help us disambiguate which kind of dependence relation underlies observed manipulability relations. For instance, mere interactions *zooming into* a phenomenon—e.g. by studying tissue under a microscope—will give us knowledge about interlevel part-whole relations or intralevel organizational features. Supplemented with this information, the dependence relations underlying Woodwardian manipulability can be analyzed and causal and constitutive relations are straightforwardly distinguished.

The current chapter has illustrated the importance of mere interactions in scientific inquiry and how they relate to other modes of scientific inquiry. The role that interventions and mere interactions play in the explanatory enterprise may be compared to the role that semantics and pragmatics play in linguistics: semantics alone will not necessarily suffice to understand the meaning of a given linguistic expression; we may need to supplement it with pragmatics, e.g. to disambiguate between different possible meanings depending on the specific context. Similarly, mere interactions can be used to disambiguate between different interpretations of manipulability observed in interventionist studies. This is not to say interventions are unilluminating. It is just to emphasize that we need more than interventions if we aim to construct scientific explanations. Yet, contemporary

---

16 Note that Silva et al. (2013) focus primarily on *integration* of evidence across various experiments. Most of these aim at uncovering causal relations by using either interventions or passive observation. When I talk about mere interactions, I refer to those kinds of experiments that are neither passive observation not interventionist studies. Also, I am not so much concerned with integration here. Rather, my focus is on classifying different types of experiments (or experimental strategies).

philosophy of science has extensively focused on interventions and neglected mere interactions. But if we want to understand how scientists explain phenomena, we must acknowledge that there is more to experimental manipulation than interventions. In chapter 14, I will start building a catalog of experiments that categorizes experiments with respect to the which kinds of manipulations they employ and which research questions they are trying to answer. While some experiments rely on mere interactions only, others will employ pseudo-interventions or combine genuine interventions with mere interactions or pseudo-interventions. This will teach us about (some of) the different manipulative strategies researchers employ to answer different research questions, about how different experiments are designed and how acquired data is evaluated. Once we understand this, we can give a more complete answer to the question of how scientists explain phenomena than contemporary philosophy of science is currently offering.

13 Excursus: A Perspectival View

Summary
Some of the difficulties I identified for contemporary philosophy of science arise from interlevel relations. Thus far, I worked with a mechanistic notion of "level" without protruding into metaphysics. Yet, understanding interlevel relations seems crucial for the business of scientific explanation. While I cannot develop a full-fledged account of levels in this book, the current chapter presents some preliminary ideas as to how levels might best be understood: as *perspectives*, i.e. different ways of accessing and describing things in the world.

13.1 Mysterious Interlevel Relations

As we have seen throughout previous chapters (especially chapters 7 and 8), contemporary "mechanisms and interventions"-style philosophy of science leaves open many questions, particularly questions concerning interlevel relations. Yet, as we have also seen, delving into metaphysics is not of great help. We do not have any convincing metaphysical account of interlevel constitutive relations; at least none that helps us tease apart constitutive from causal relations based on empirical evidence (see also chapter 7). Also, knowing the precise metaphysical underpinnings of constitutive relations does not help us understand how scientists cope with the underdetermination in their empirical data. Therefore, in order to see how scientists come up with explanations we need to examine more closely the experiments they carry out, their manipulative strategies and experimental designs as well as their methods and data analysis.

I have argued in chapter 11 that (IV_{dm})-defined interventions make it possible to marry interventionism and mechanistic explanations. As discussed in chapter 12, taking on board non-intervention experiments, viz. experiments using mere interactions rather than Woodward-style interventions, fills a gap in theorizing about experimental manipulations. It gives us a means to discover, e.g., the geometrical or *architectural features* of a system including what parts it consists of, what their size, shape, and arrangement is, how they are connected, and so forth. It thus gives us a means to open up black boxes, "zoom into" a phenomenon, and identify a starting point for applying interventions (in the sense of (IV_{dm})-defined interventions).

However, the question remains open as to how interlevel relations are best understood. Mechanists talk about *constitutive* part-whole relations, about *making up, production*, and *bringing about* to describe these relations. They urge that levels

DOI 10.1515/9783110530940-013

are relative to a given mechanism. This raises at least two issues I have pointed out already. First, we do not know what it means for two entities to be on the same level (see chapter 3.7). Second, if we cannot manipulate multiple levels independently due to their non-causal dependence, it is difficult to see in what sense a given manipulation can be said to be on a certain level (see chapter 6.4). One way to respond to these issues is to abandon levels from the neuro- and life sciences altogether and speak of composition or scale instead (e.g. Eronen 2013, 2015). I believe, however, that the levels metaphor captures something important that the notions of scale and composition do not capture, viz. certain epistemological features. Therefore, the current chapter suggests an understanding of levels as *perspectives researchers take*. This idea is *not* fueled by *a priori* metaphysics but by empirical research practices. If we conceive of levels as perspectives, what constitutes any given level and how it relates to other levels essentially depends on the tools and methodologies scientists use to acquire and evaluate data. This view is ontologically sparse, empirically realistic, and inspired by a classic position in philosophy of mind: the identity theory.[1]

13.2 Place's Idea

The *identity theory* (IT) is a theory about how mental and physical processes relate. It is a traditional view in the philosophy of mind, specifically in the context of debates about reduction of the mental to the physical. The classic textbook version of IT defends the view that mental processes are *a posteriori* identical to physical processes (e.g. Beckermann 2001).[2] Accordingly—to cite an overused and biologically inaccurate example—pain just *is* c-fiber firing. There is nothing to the mental "over and above" or "outside" the physical; and this is not a matter of definition, it

---

1 Note that Craver (2013) himself advocates a "perspectivalist view" of mechanisms, though he means something different than I do. According to Craver's perspectivalism, there are three aspects of mechanistic explanations: constitutive, contextual, and etiological. While I do not see any reason to disagree with this, my project here is a different one. I am interested in understanding the constitutive interlevel aspect of mechanistic explanations and I suggest conceptualizing interlevel relations as relations of different perspectives a scientist can take. The perspectival view I develop in this chapter bears certain resemblances to Giere's (2006) *scientific perspectivism* as well as Wimsatt's (1976; 2007) view of levels. Unfortunately, a detailed discussion of these views is beyond the scope of this book. For current purposes it shall suffice to sketch the idea of levels as perspectives in broad strokes and leave a detailed account for another occasion (see Kästner in press).

2 It should be noted that in the current context I am concerned with type IT, not token IT, as we are discussing general, type-level scientific explanations.

is a matter of fact. Thus, although the expressions "being in pain" and "being in brain state *b*" are not synonymous (they express a different meaning or Fregean *sense*) we know that if it is true that being in pain is being in brain state *b* it will be true that whenever a person is in brain state *b* she is in pain and vice versa.

Attracted by metaphysical simplicity and parsimony, for instance John J.C. Smart (1959) argued that science will eventually be able to account for even complex higher-level phenomena such as consciousness or the behavior of humans in terms of purely physical processes:

> It seems to me that science is increasingly giving us a viewpoint whereby organisms are able to be seen as physico-chemical mechanisms: it seems that even the behavior of man himself will one day be explicable in mechanistic terms. There does seem to be, so far as science is concerned, nothing in the world but increasingly complex arrangements of physical constituents. (Smart 1959, p. 142)[3]

Another virtue of identifying mental with physical processes is at hand: we cannot run into problems of causal exclusion such as the one discussed by Kim (see chapter 4.2.1). However, there are also drawbacks. Most prominently discussed are perhaps concerns about multiple realizability. As Hilary Putnam (1975 [1967]) famously pointed out, it seems highly unlikely that different species—mammals, reptiles and even aliens—would need to have exactly the same brain state in order to feel pain; maybe they do not even have a central nervous system like we do. Rather than identical to brain states, pain should be viewed as *multiply realized*, viz. realized by different physical substrates in different species.[4]

Admittedly, there have been many different versions of IT and there has been considerable debate as to how strongly to conceive of the identity claim. This is not the place to enter into these debates.[5] Yet, I side with Polger and Shapiro in recognizing that

---

3 Note that despite the proposed identity Smart talked about different *viewpoints* and *seeings* of complex physico-chemical mechanisms. I will get back to this below.

4 Historically, arguments from multiple realizability lead to a rejection of IT and its replacement with *functionalism*, viz. the view that types of mental states are individuated by the role they play. But IT has been revived in more recent debates about supervenience and phenomenal consciousness (e.g. Kim 1998; Pauen & Stephan 2002). One proposal to meet challenges from multiple realization has been that we can constrain where identity statements are applicable. For instance, we may restrict the identification of pain with c-fiber firing to healthy adult human beings (Newen 2013 makes this point explicit).

5 See van Riel (2014, ch. 3) for a detailed discussion of the identity theory and reduction in the philosophy of mind and how it can be linked to debates in philosophy of science.

[w]hen rereading the original papers of Place, Smart, and Feigl, one cannot help but suspect that the canonical interpretation of their ideas has been unfair. Far from being a slow and wounded stag that the arrow of multiple realization might easily bring down, the identity theory, as originally presented, proposed a versatile and prescient conception of the relation between the mind and the brain. (Shapiro & Polger 2012, p. 270)

In a seminal paper, Ullin T. Place (1956) discusses two types of identity statements. Statements like "a square is an equilateral rectangle" are true by definition; whenever something is a square, it is also a equilateral rectangle.[6] But there are also statements like "this table is an old packing case" which are true only because there happens to be an object that is both, a table and an old packing case. Although the expressions (being a table, being a packing case) seem quite unconnected, they can both provide an adequate description of the same object in particular circumstances (when there is an old packing case that is being used as a table). Place calls the "is" in such statements an *"is" of composition*. He illustrates the relation expressed by such an "is" of composition using the example of a cloud:[7]

A cloud is a large semi-transparent mass with fleecy texture suspended in the atmosphere whose shape is subject to continual and kaleidoscopic change. When observed a close quarters, however, it is found to consist of a mass of tiny particles, usually water droplets, in continuous motion. On the basis of this second observation we conclude that a cloud is a mass of tiny particles and nothing else. But there is no logical connexion in our language between a cloud and a mass of tiny particles; [...] the terms 'cloud' and 'mass of tiny particles in suspension' mean quite different things. Yet we do not conclude from this that there are two things, the mass of particles in suspension and the cloud. (Place 1956, pp. 46–47)

The identity expressed by the statement "a cloud is a mass of tiny particles in suspension" parallels, according to Place, identities between mental processes and processes in the brain. Such statements combine observations made about a single object with different tools or from different perspectives that might be mutually exclusive:

We can observe the micro-structure of a cloud only when we are enveloped by it, a condition which effectively prevents us from observing those characteristics which from a distance lead us to describe it as a cloud. (Place 1956, p. 47)

Analogously, psychologists may observe mental processes in a subject while conducting an experiment (say, by using some kind of clever questionnaire) while

---

6 Place calls the "is" in such statements an *"is" of definition*.
7 Note that Place's "is" of composition view is quite different from the coreference view mentioned above.

neuroscientists can observe neural processes in the subject's brain using some fancy neuroimaging technique. Although their observations are quite different and each of their tools would not permit to see what is accessible through the tool the other is using, they are actually making observations about the same thing: the same (neural/mental) process in the same subject.

This kind of identification is substantially different from what the classical textbook version of IT suggests: far from there being a rigid one-to-one mapping between physical and mental processes, there seems to be a continuum of different observations. This suggestion is not unique to Place's work. Indeed, Smart talked about different *viewpoints* or *seeings* of complex things, too. It is the relation between such different observations (viewpoints, perspectives, or seeings) of the very same process which is expressed by Place's "is" of composition. The identification Place suggests (and Smart would probably agree) is not bound to there being just one way in which things can be composed. For instance, he argues, there is nothing self-contradictory about claiming that while one cloud is a mass of tiny particles another is a cluster of bits of fabric. Thus, multiple realization is not a threat to identification.

The crucial question now is under which conditions we can apply the "is" of composition to identify different observations and attribute them to the same process rather than two different processes. In the case of the cloud this is easy: look at it from the valley and climb a mountain until you are in the middle of it and you will know the same thing from different perspectives. But how about cases where we cannot do that? Clearly, Place holds, the observations must be correlated. However, this is not enough; for processes related as cause and effect are also systematically correlated, yet we would not want to identify, say, the position of the moon in relation to the earth with the tides.[8] In an attempt to disentangle observations related by causation from those related by composition Place suggests we should treat two observations

> as observations of the same event, in those cases where the technical scientific observations [...] in the context of the appropriate body of scientific theory provide an immediate explanation of the observations made by the man on the street. (Place 1956, p. 48)

---

8 Place also discusses that it is characteristic of statements that can be related by an "is" of composition that they describe observations which cannot be made simultaneously. Just like we cannot see the could while we are surrounded by the particles it is made of, we cannot see the mental processes when we are looking at individual neural signals, etc. However, just because we cannot see one thing while we see another does not mean the two can be identified. I may not be able to see Robert and Helga at the same time because they dislike one another too much to ever occur in the same place at the same time. But that does not lead me to think Robert is Helga.

But when does one observation provide an *immediate explanation* for another? Though Place does not discuss this question in great detail he explicates the explanatory relation in terms of a common causal history leading up to both the explanans and the explanandum observation. There is effectively just one thing that is happening, observed in two different ways. The perspectival view I am going to suggest resembles Place's IT. And just like Place's original suggestion, it is quite different from textbook versions of IT.

13.3 A Matter of Perspectives

The basic idea of the perspectival view can be illustrated in terms of Place's cloud example. There is only one cloud, but just as we can view it from a distance as a meteorological phenomenon we can examine it from within to study its composition. We can zoom into and out of the cloud by choosing different perspectives, and reveal different features of the cloud by using different methodologies. A chemist, for instance can analyze the precise molecular structure of the water droplets and how they are arranged. A meteorologist can analyze how the cloud as a whole moves in the air, and so on. Likewise, we can study cognitive processes from a psychological point of view or by taking functional brain images. Still, just as when examining the cloud, we are observing the same thing, viz. a given cognitive phenomenon.[9]

But how do we know that different observations are actually observations of the same thing? Place suggests we can infer this whenever the observations in question are correlated and share a common causal history. This suggestion matches what we see in scientific practice: scientists often build their explanations on correlated observations; this is especially true for the kinds of interlevel (constitutive) mechanistic explanations that try to account for cognitive phenomena in terms of neural processes. I have outlined multiple examples in chapters 2.3 and 7 where data about cognitive processing (e.g. memory performance) is *correlated with* neuroimaging data. In the experimental setting, scientists assume that they are studying 'the same' phenomenon because they are asking participants to perform a specific task (e.g. a recognition task) and take both performance and neural activation measurements simultaneously. This mirrors Place's requirement for a common causal history and correlated observations.[10]

---

9 As I do not intend to enter into metaphysical debates, I will work with a non-technical notion of phenomena.
10 This is not to say, however, that all and only correlated observations are observations of the same phenomenon. Perhaps, we want to relate EEG data to fMRI data but cannot apply both

The principle suggestion that observations at different levels might be conceived of as (correlated) observations from different points of view is not unfamiliar from the mechanistic literature. Just as we can zoom in and out of Place's cloud we can look at a mosaic (to use Craver's picture) from a distance or examine how the different pieces are fitted together. Similarly, we can look at a car as a whole (to use Glennan's picture) or get a mechanic to look under its hood. Either way, what we see depends on the perspective we take (which tools we use) and we might not see the bigger picture when examining the details of a sub-structure it contains. Likewise, we may loose details when just describing something at a higher-level, using coarser-grained generalizations. Coarser-grained generalizations might be less stable than those describing processes in a more fine-grained way. This is because higher-level generalizations may have to accommodate for quite some variability at lower levels (see also Woodward 2013 on this) which may e.g. come from multiple realization.[11]

My suspicion is that a many-layered view of the world so often advocated is merely a result of descriptive diversity, which in turn is rooted in our investigating phenomena using various different methodologies, i.e. observing phenomena from various different perspectives. Since different methodologies have different strengths and weaknesses (see figure 13.1), they naturally permit examination of different features of a phenomenon. This takes me to a second point. It is not only that different tools (or perspectives) allow us to study a phenomenon at different resolutions, different *levels of grain* if you will. Just as using different methods allows scientists to zoom and in out of the phenomenon, they may also study different portions of it. For illustration suppose the eastern portion of Place's cloud is hanging over a mountain top while the western portion is hovering over a valley. Climbing up the mountain will give us access only to the eastern but not the western portion. Analogous scenarios can be found in neuroscience. Say, for instance, we are looking at the molecular processes during action potential propagation. If we somehow mark sodium molecules we can visualize sodium influx while we may not see potassium efflux. To see this, we would have to apply a different tool, viz. mark potassium molecules. Still, sodium influx and potassium efflux are both present during each action potential generation. But it takes different tools to see them.

---

measurements together. In this case, a clever experimental design must be found that "reproduces" the phenomenon in question in both an fMRI and an EEG study.

11 Note that this view—though arguably apt in paradigmatic mechanistic cases—is not uncontroversial. For instance, Potochnik (2010) discusses cases from population biology to argue that we cannot simply assume higher-level explanations to be more general than lower-level ones, or even genreal in the right way.

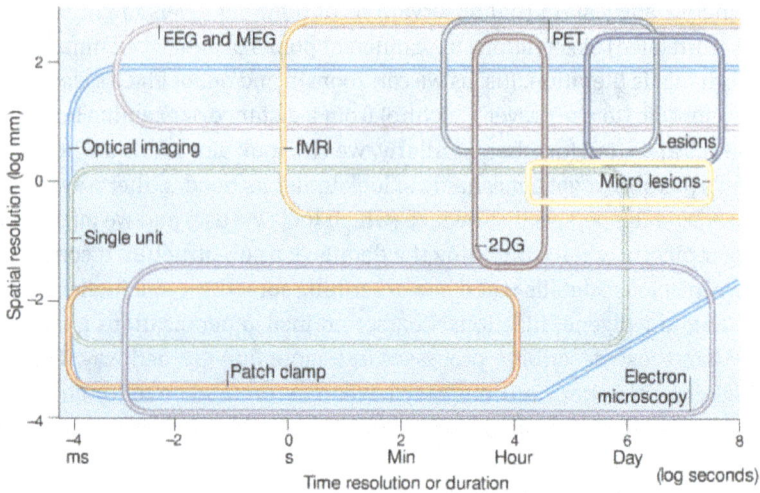

Fig. 13.1: Methods in cognitive neuroscience; schematic illustration of their temporal and spatial resolutions. Reprinted by permission from Macmillan Publishers Ltd: Nature Reviews Neuroscience (Grinvald & Hildesheim 2004), copyright 2004.

To make matters even more complex, scientific (correlated) observations may also be observations of different portions of a phenomenon that are also observations at different resolutions. The memory research discussed in chapter 2.3 is a case in point. For instance, we can access individual neurons in hippocampus in model organisms, but we do not have this kind of information about other (higher) cortical areas such as prefrontal cortex. Although we are still studying the same phenomenon, the different perspectives we take do not have to neatly cut up a phenomenon into chunks (like windows aligned in a row) or hierarchical slices (so that we can zoom in and out); though this is in principle possible, of course.

In this sense different scientific tools and methodologies provide different windows into the world. Each of these windows shows different portions of the world, lets through different amounts of light, and distorts the appearance of the world out there in a different way. If we are aiming to understand and explain something, what we need to do is not just stare out of one window but hobble along from one window to the next and combine what we see into a bigger picture, correcting for different distortions as we go along, inferring something about the bits we cannot see, and perhaps even try to polish our windows (i.e. improve our methodologies) and discover new ones as we go along.

What mechanism (or parts of a mechanism) and what features thereof we see essentially depends on the perspective we take, viz. the method we use to investigate a phenomenon, the window through which we look at it. Figure 13.2 illustrates this. Importantly, once again, levels so understood do not necessarily form exact part-whole hierarchies. They cannot necessarily be stacked neatly onto one another like putting together a Matryoshka doll (though this is still possible). They also do not necessarily relate by supervenience (see Potochnik 2010). Generally, different levels may pick out different features at different levels of grain. The resulting mechanisms we observe may thus overlap or intersect just as well as they might be independent or related in a part-whole fashion. It is a matter for scientists to find out what the precise relation between observations at different levels—i.e. observations made using different tools or taking different perspectives—and the mechanisms they reveal is. This is not a task philosophers can solve from their (metaphysical) armchairs.

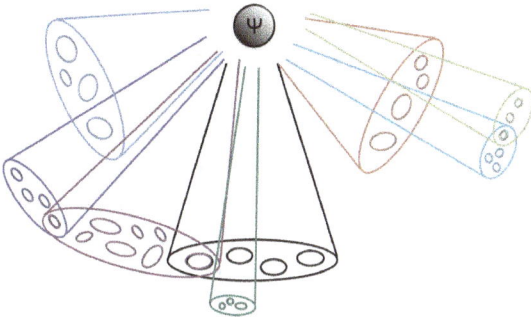

Fig. 13.2: Scientists uncover different mechanisms when studying a phenomenon (Ψ) using different tools; the different mechanisms (pictured in different colors) they see might be related as part and whole, independent, or partially overlapping.

13.4 Lessons Learned

Although the perspectival view leaves open the metaphysical details of interlevel relations, it is intuitively appealing. It preserves descriptive diversity while it is ontologically sparse. No more than a single, unified, say physical level of real-world entities is needed.[12] Yet, our explanations may well make reference to more complex, higher-level, phenomena. By scientists this view should be welcomed because it reflects the deep conviction of many that "in some fundamental sense the mind just *is* the brain" (Bermúdez 2010, p. 6, emphasis added) while it leaves room for different descriptions (e.g. cognitive and neural ones). By philosophers it should be welcomed because ontological unity forecloses worries about causal preemption, interlevel causation, emergence of spooky higher-level properties and so forth.

Proponents of mechanistic explanations should be particularly happy with the perspectival view. First, it mirrors a key feature all advocates of mechanistic explanations share: mechanisms are always mechanisms *for something*, and insofar mechanistic explanations are always interest-relative (cf. Glennan 1996, 2010b; Machamer et al. 2000; Craver 2007b, 2013; Illari & Williamson 2012). Moreover, it is not just that there are different mechanisms for different phenomena but also different mechanisms depending on how a given phenomenon gets studied. This, in turn, may depend on the particular research question we are asking. For instance, if we want to know whether the key will sink when dropped into the water it will be relevant whether it is made of heavy iron or light plastic; but its exact shape is irrelevant to this question. But if we wonder whether the key will open a given lock, this information will not help; for answering this question needs information about its shape and rigidity. Answering these questions requires examination of different features: investigating lower-level or micro-structural features will illuminate what material the key is made of while studying higher-level or macro-structural features will reveal its shape and rigidity. But still, there is only one key, though we can describe it "at different levels" and descriptions at different levels are relevant to answer different research questions.[13] However, some mechanists may object that the perspectival conception of "levels" does not respect the mechanists' view of levels as strictly relative to any given mechanism (Craver 2015). But, as outlined in chapter 3.7, this view is problematic anyway: either we cannot identify interlevel causal relations because we cannot know what is on the same level, or we use causal relations between entities to infer that they are on the same level and loose the strict relativity mechanists advocate.

Second (and relatedly), on the perspectival view we gain an understanding of what it means for two things to be on the same level, viz. being *accessible through the same method* (or the same set of methods). While the general idea is

---

12 The principle idea the perspectival view advocates does not hinge upon postulating a single-layered purely physical world. However, I take this to be the least controversial, metaphysically least demanding position.

13 This example is due to Glennan (2010b).

that different methods target different levels, I do not suggest that there cannot be multiple methods targeting the same level. What is and what is not at the same level is most likely a question for empirical scientist to answer. It is true that, in a sense, I trade one problem for another here. The perspectival view tells us what is at the same level (in mechanisms); but to do this, we need an account of what makes for the *same method* or *same intervention*. However, I think that this latter question—unlike the original one—will have a more or less straightforward answer given what we see in empirical research. The very starting point of my project is that scientists (at least sometimes) successfully come up with explanations. Thus, we can assume that scientists (at least sometimes) get it right what counts as "the same method" or as "the same intervention" (and therefore as "on the same level"). Examining how scientists gain evidence through different methodologies and integrate it will thus be illuminating with respect to the question what counts as the same method, intervention, or level. Unfortunately, a detailed analysis of experimental research providing concrete criteria for what counts as the same method, intervention, or level is beyond the scope of this book. However, it is quite clear already that once we adopt the perspectival view, we can answer this question based on empirical practice rather than having to delve in *a priori* metaphysics. This is a virtue of the perspectival view; it renders it empirically plausible and avoids heavy metaphysical burdens.

Once we know what counts as accessible though the same method, we gain a simple handle on the problem Fazekas & Kertész (2011) raised for mechanistic explanations (see chapter 3.7). Put briefly, their argument was that if we have no way of finding that two mechanisms are "at the same level", but insist that causal relations obtain intralevel only, we cannot piece the mosaic together. Now adopting the perspectival view offers a straightforward solution: if we can access two mechanisms with the same kind of method, they are at the same level.[14] We can thus integrate experimental evidence across individual observations, even when those were obtained in different places, at different times, using different individuals, perhaps even different experimental designs. Such integration of evidence across multiple experiments is part and parcel of contemporary cognitive neuroscience (see chapter 2.3). It is vital if we want to hold onto some kind of

---

[14] Since the perspectival notion of levels is much weaker than a part-whole relation, interlevel relations on the perspectival view are to be understood much looser than mechanistic constitution. They might be a part-whole relations, but they might also be something else. To find out what exactly they are, though, is not a task for philosophers. It is something for scientists to figure out. Sometimes we clearly work with part-whole relations. Everywhere where we can look into something or take it apart, we effectively decompose a whole into its parts. Such studies are typically cases of mere interactions (see chapters 12 and 14).

mutual manipulability criterion to establish mechanistic componency but cannot perform bottom-up and top-down manipulations in a single experiment.

Third, the perspectival view illuminates what mechanists mean by *production*. Basically, there seem to be two ways to conceive of production (also referred to as *productivity*, *exhibiting*, or *bringing about*). On the one hand there is a causal reading of these terms according to which production is an—intralevel—matter of physical causation or mark transmission (e.g. Salmon 1984; Glennan 2010b). On the other, there is a non-causal reading of production which is at play, e.g., when we talk about neural activations *producing* cognitive phenomena. On this reading, productions are interlevel relations like realization or implementation.[15] Realizers produce the phenomena they realize (or implement). A given cognitive process is produced by (realized by, implemented by) certain neural processes. This kind of relation is sometimes also referred to as a *making up* or *building* relation holding between the components together realizing/implementing/producing a phenomenon and the phenomenon as a whole. It is what Place called "composition".[16] Insofar as there are multiple ways to realize, make up, compose, produce, implement, or build a phenomenon we may speak of multiple realization or *alternative constitution* (cf. Ylikoski 2013).

Used this way, the notion of "production" may remind of old debates about *emergence*, viz. how complex (higher-level) processes result from an accumulation of relatively simple (lower-level) processes.[17] Without delving into detailed debates about emergence, it is interesting to note that emergence is usually considered somewhat spooky in philosophy. Yet, it is well-respected in other sciences such as computer science and artificial intelligence. In these disciplines, the emergence of *complex patterns* as a result of iterated relatively simple processes is part and parcel of system design and analysis; there is nothing spooky or mysterious about it. Patterns are simply made up from their underlying processes just as clouds are composed of water droplets. Talking about a pattern rather than all the underlying processes is effectively taking a different perspective; it is looking from a distance

---

15 For a discussion of "realization" in the context of interlevel scientific explanations see e.g. Polger & Shapiro (2008), Polger (2010), and Gillett (2010; 2011).

16 In places it seems that this "making up" is used interchangeably with "constitution". However, from the discussions so far I would conclude that a constitutive relation holds between a phenomenon as whole and an *individual* component while a making up or building relation is a relation between assembled components and the phenomenon as whole. For more on constitutive relations see also Bennett (2011) and Ylikoski (2013).

17 Emergence has been discussed both in the context of mechanical philosophy (e.g. Broad 1925) and D-N explanations (Hempel & Oppenheim 1948). For an overview on different conceptions of emergence see Stephan (1999).

rather than examining minute details.[18] And sometimes this is much more telling and/or much more efficient than examining the underlying processes in detail (see also Dennett 1991). Just as computer scientists do not have to be afraid of patterns being something spooky, I suggest, philosophers of science need not fear emergence. Seeing emergent phenomena can be as simple as "zooming out" of its underlying (realizing, implementing) processes. It is a matter of taking a look at the bigger picture, using less magnification. If emergence is understood in this way, it is thoroughly unexciting and philosophers of science need not fear it. This, in turn, illuminates how explanations involving complex (dynamical) systems relate to decomposable mechanisms with clearly localized parts (an issue I raised in chapter 3.7): they are essentially different descriptions (or models) of the same thing. We can describe one thing in multiple different ways as we look at it from different perspectives. And each perspective has its own advantages. Although such different perspectives might in principle be independent, we should expect them to eventually constrain one another when we start piecing together the mosaic.

Fourth, we now understand how interlevel experiments are different from intralevel ones: interlevel experiments combine methodologies at different levels while intralevel ones do not. Further, we can address the question of how to place interlevel manipulations onto a specific level even though the manipulation will affect multiple levels at once.[19] It targets only one level because it is applied from a given perspective (using a given method). This squares well with the consensus view among mechanists that interlevel relations cannot be causal because they do not relate distinct entities (cf. Machamer et al. 2000; Craver & Bechtel 2007; Craver 2007b,a; Glennan 2010b; Illari & Williamson 2012; Woodward 2013). For if different levels simply offer different perspectives on the same thing, interlevel causation would lead to the odd conclusion that this one thing causes itself.[20] Once we know if our manipulations are interlevel or intralevel in nature, we can disambiguate observed dependence relations even in the absence of separate conceptions of interlevel and intralevel manipulability. This squares well with adopting (M_{dm})-(IV_{dm})-interventionism.

---

18 For more on the relation between patterns and mechanisms see Kästner & Haueis (2017).

19 We encountered this problem in chapter 6 when discussing the common cause interpretation of interlevel experiments.

20 There might be some view of causation according to which this is possible but I am here working with a standard conception of causation according to which causes produce their effects, not themselves, and effects do not bring about their causes.

13.5 Worries

Before closing, I want to briefly address two objections readers might raise against the perspectival view. The first is that the perspectival view is in fact a disguise for reductionism, the second is that it permits an inflationary explanatory pluralism. I shall take these in turn.

On some readings of "reduction" it may very well be the case that the perspectival view is reductive in nature (see van Riel (2014) for a detailed discussion of different conceptions of reduction). But I am not concerned about such nomenclature. The important point is that I do *not* intend to support views according to which higher-level processes can be simply replaced with lower-level ones. The "ruthless" reductionist view advocated by Bickle and colleagues (Bickle 1998, 2006; Silva et al. 2013) is a case in point (see chapter 3.6). Far from being dispensable, I argue that higher-level entities, processes, and events have their place in science and scientific explanations. And they are not just heuristics or black boxes to be opened; they may do some genuine explanatory work. Moreover, if the different perspectives we take overlap, we cannot demand a straight one-to-one mapping of different levels onto one another. This is not to say that no such mappings exist (in ideal cases they probably will), but that we should not generally expect them to. If you say this is reduction then the perspectival view is a reductive view. But it is important to note that this kind of reduction is not what mechanists struggle against (see chapter 3.6).

How about the explanatory pluralism the perspectival view suggests? Some may worry that it is much too simple and licenses an inflationary amount of different, perhaps competing, models to explain a single phenomenon. But when scientists are seeking to explain a phenomenon, they do aim to confirm or disconfirm various models to find the "best" one (irrespectively of what precisely makes for the "best" explanation), after all. Without comparing and judging different models and explanations, it will be hard to make progress in science.

This worry is legitimate. However, I do *not* say that just because there are many different perspectives we can take on a phenomenon, these different perspectives, the resulting models and explanations, cannot disagree; they can. For instance, if two research groups use the same principle method (or combinations of methods) and similar paradigms to assess the same phenomenon we should expect them to agree. This is because they are basically taking the same perspective, they look through the same window, so they should have access to the same features of the explanandum phenomenon. Thus, their explanations should in principle be in agreement.[21] Additionally, I take it, the explanations we come up with on each individual level will mutually constrain one another as we try to link the different domains. Discarding what does not fit the overall picture or revising the overall picture to fit new evidence we want to hold onto will give us room to maneuver our way between all the possible explanations towards what we may think of as the best one (for whatever reason) at the time. Therefore, the kind of explanatory pluralism the perspectival view suggests is nonhazardous.

13.6 Upshot

All in all, the perspectival view illuminates a promising path towards understanding how to think about "levels" in the context of scientific explanations. It is metaphysically non-committal, empirically realistic, acknowledges that experimental observation uses a variety of strategies, permits multiple realization, and leaves the identification and precise relations between different levels a matter for empirical scientists to discover. [22][23]

There seems to be a consensus among defenders of all the different brands of mechanistic explanations that what is needed to come up with an explanation is (i) identification of the explanandum phenomenon, (ii) its decomposition into relevant entities and activities (and perhaps also identifying relevant background conditions), and (iii) analyzing how they produce the explanandum phenomenon, i.e. how they must be organized to be responsible for the characteristic behavior of the mechanism as a whole (see Darden 2006; Bechtel & Abrahamsen 2008; Craver 2013; Illari & Williamson 2012). And indeed this view is much older than recent mechanistic theories. Already in the late 1980s Marvin Minsky recognized that there is a principled three-step strategy

> for understanding any large and complex thing. First, we must know how each separate part works. Second, we must know how each part interacts with those to which it is connected.

---

21 They might still have different interests; say, one seeks a particularly accurate model, the other a computationally tractable one. In this case, the explanations will still be different, but this is not a case of genuine disagreement for both parties know their models are tailored to fit different purposes.

22 If I was forced to commit to any ontology, I would pick a purely physicalist one; I take it that all I say is compatible with physicalism. Yet, I do not think such a commitment is necessary in the current context.

23 It is certainly a valuable project to spell out the metaphysics of mechanisms, but this is beyond the scope of this book (however, see Krickel 2014). Whatever the suggestion will be, I take it that it should respect what we learn about levels and interlevel relations from scientific practice.

And third, we have to understand how all these local interactions combine to accomplish what the system *does*—as seen from the outside. In the case of the human brain, it will take a long time to solve these three kinds of problems. (Minsky 1988, p. 25)

The basic idea of the perspectival view is that all of this can be achieved by studying a phenomenon from different perspectives, using different observational and manipulative methods, effectively zooming in and out of the phenomenon and examining different parts of it. To fully understand how scientists come to explain (cognitive) phenomena, we need to learn about the precise structure of the experiments they conduct and how different manipulations get systematically combined. In order to do this, developing a catalog of experiments is in order. This will be the agenda for the final chapter.

14 Mere Interactions at Work: A Catalog of Experiments

Summary

Reassessing some of the examples from chapter 2, this chapter will illustrate how different manipulations get used in scientific practice. A particular focus will be on experiments using mere interactions rather than Woodwardian interventions and mutual manipulation. Studies combining different kinds of manipulations (including mere interactions, pseudo-interventions, and interventions) will also be discussed. Over the course of this chapter, I will develop a catalog of experiments into which scientific studies can be classified according to the kind of research question they are investigating and the experimental manipulations they employ to this end. This is a project in *philosophy of experimentation*; it will help us gain a deeper understanding of how different kinds of explanatory information are acquired experimentally and how evidence from one experiment can be fed into designing subsequent ones.

14.1 Philosophy of Experimentation

Thus far I have argued that contemporary "mechanisms and interventions"-style philosophy of science faces some severe challenges. The problems related to marrying interventionism with mechanistic explanations can be cleared away by adopting (M_{dm})-(IV_{dm})-interventionism and a perspectival view on levels. However, a crucial problem remains: the fact that focusing on Woodward-style interventions is too short-sighted and leaves open how scientists come to make those assumptions they need to make to get their intervention studies going. To successfully carry out intervention experiments, scientists need to know how and where to apply the desired interventions. Further, to disambiguate which kind of dependence underlies observed manipulability, scientists may need to know something about the architecture of the system (e.g. where part-whole relations obtain). Besides, although it is quite true that explanations are often based on manipulative experiments, manipulations are not exhausted by interventions. Indeed, scientists typically employ different kinds of manipulations as well as different methodologies to study the same phenomenon. If we are trying to understand how explanations of cognitive phenomena are constructed by cognitive neuroscientists, we need to understand what precisely it is that they are doing. That is, we need to understand what different experimental strategies they use, which research questions each can answer, and how different manipulations get combined. This is what I call

DOI 10.1515/9783110530940-014

philosophy of experimentation. It requires close consideration of the *materials and methods* scientists employ in empirical research.

Notice that what I am after here is *not* to delve into the vices and virtues of different experimental tools as such, how to employ them, and what their practical limitations are; this makes for a much bigger project.[1] For now, I am talking about the logic behind experimental manipulations, experimental strategies or types of *experimental designs* rather independently of the differences between different methodologies (say, EEG and fMRI). Of course, it would be foolish to deny that there are mutual constraints. But this is not my topic here. Rather than examining the concrete experimental tools or techniques, their vices and virtues and practical limitations, I want to shed light on the principled ways in which experiments are designed and the different ways in which explanations can be inferred from the acquired data.

I have argued in chapter 12 that there are at least some empirical studies involving manipulations which do not qualify as Woodwardian interventions. Having established that these *mere interactions* are interestingly different from interventions, yet important and frequent in scientific research, we may wonder how exactly such mere interaction studies get used in cognitive science.[2] This we can learn by examining experimental practice in more detail. In what follows below I will sketch different types of experiments all relying—to a certain extend—on mere interactions. Though I do not claim the list to be complete, it clearly illustrates that there is more to scientific experimentation than interventions and that some of these non-interventionist manipulations can be used to discover the very thing scientists aim to intervene into in subsequent experiments.

The experiments I will be discussing are those kinds of experiments that exist in addition to—or are behind some of—the clean bottom-up and top-down interventions described in chapter 2.3. I will return to some of the examples discussed there in due course and illustrate that they are not in fact as clearly interventionist-style experiments as the success stories told in neuroscience textbooks make us believe. As we will see, some mere interactions just serve to systematically acquire data, some ensure appropriate contrasts or background conditions, or that an intervention worked in the desired way, and some are employed to create different experimental groups and conditions. Where different groups are compared, we sometimes find *pseudo-interventions*, viz. manipulations that are interpreted

---

1 But see Senior et al. (2006) for a discussion on the most commonly used empirical methods in cognitive neuroscience.

2 Indeed, contra Silva et al. (2013), I think that there rarely is empirical research without any interaction at all. For any kind of data acquisition will require at least a minimal amount of interaction with the phenomenon or system under investigation.

as if there had been genuine Woodward-style interventions although there have not been (see also chapter 12). Despite the fact that such manipulations are not proper interventions, they systematically do make valuable contributions to the scientific enterprise—either independently or in combination with Woodwardian interventions. Therefore, if we aim to understand how scientists explain (cognitive) phenomena, we better not neglect any of these strategies.

I systematically examined a range of experimental approaches we find in cognitive neuroscience. Classifying these studies according to which manipulations they employ and which research questions they are trying to answer yields the following (preliminary) catalog of experiments.

14.2 Geometric Property Studies

Recall the staining example introduced in chapter 12.2. I argued that if we stain a tissue sample to reveal aspects of the cellular architecture we are not intervening into the sample with respect to its structure but merely interacting with it. This particular mere interaction belongs to experiments I shall call *geometric property studies*. Geometric property studies reveal structural aspects of a system through physically manipulating, analyzing, or decomposing it. In a way, these experiments are the very prototype of mere interaction studies that put things under a metaphorical magnifying glass. Geometric property studies can be carried out at different scales or levels (e.g. the whole organism or the single cell). They are primarily concerned with the geometric properties of entities, viz. their size, shape, and arrangement, including their position, location, orientation, and connections. While they may require some prior knowledge as to what qualifies as an entity or fits an established category (say, for instance, what is a neuron and what is not), they typically bring out aspects of the system or phenomenon in question that are inaccessible to us without using any kind of manipulation.[3]

Further examples of geometric property studies include different kinds of imaging (fMRI, x-ray, DTI, x-ray crystallography, etc.), slicing, preparing samples for histological analysis, or otherwise or decomposing a system into parts. For illustration consider Korbinian Brodmann's (1868–1918) work. Based on *cytoarchitectonics* (i.e. the structure and organization of cells), Brodmann mapped human cortex in 52 different areas—known as *Brodmann areas*. Brodmann's atlas has since become

---

3 Different geometric property studies might only be available and illuminating in specific circumstances or given specific research questions. However, this does not diminish their role in cognitive neuroscience as well as the empirical research more generally. Mere interactions employed for studying geometric properties are all over the place.

a major point of reference for neuroscientists. Indeed, many of the areas Brodmann identified have meanwhile been associated with specific cognitive functions—but Brodmann's work was purely histological. He interacted with the tissue he was studying in various ways—slicing it, staining cells, putting it under the microscope, and so on—but he did not change anything about it *in order to observe the effect of his manipulation*. He did not carry out Woodward-style interventions, that is. The same principle point can be made once we reconsider the works by Felleman and van Essen (1991) discussed in chapter 2. Primarily by examining the architecture of visual processing areas, their histology and connections, Felleman and van Essen identified 32 functionally different brain areas responsible for visual processing and more than 300 connections between them.

Although such manipulations make accessible new and valuable information, they do not qualify as Woodwardian interventions in the current context. It is not the fact *that* some cells are stained that makes them an important contribution to the explanatory enterprise, but what we see once the cells have become stained. What is illuminating is not that the cells get stained but the structure that is revealed as a result. This is why, although clearly not an interventionist manipulation, stains have significantly informed neuroscience.[4]

The attentive reader may object that maybe putting samples under the microscope, or collecting structural brain images, does not change the tissue, but that slicing and staining it certainly does. This is true. But even so, as I have argued in chapter 12.2, these are not manipulations in the interventionist's sense. I do not doubt that Brodmann or Felleman and van Essen manipulated the tissue they worked with, but the question is *with respect to what* did they manipulate it? They did not slice up a brain to see what happens when you slice up a brain. Neither did they stain cells to see what happens if you apply a stain. This they already knew; and this is why they purposefully employed these manipulations to reveal something else, viz. the cytoarchitectonics. In this context, slicing up a brain or applying a cell stain is not very different from putting some tissue under a microscope.[5]

---

4 Spanish neuroanatomist Santiago Ramon y Cajal (1852–1934) famously employed a modified version of Golgi's technique to discover organizational features of the nervous system that later evolved into the *neuron doctrine*, viz. the conviction that the nervous system is built from individual cells—neurons.

5 There is another sense in which cell staining is still slightly different from slicing. While slices show things that we could in principle see if our eyes would penetrate the whole, stains reveal structural aspects that are hidden from our view without them, even in the thinnest slices. This difference, however, does not mark the difference between intervention and mere interaction. Rather, it shows that different tools can be put to use when we merely interact, even when we are just looking at geometric property studies.

The same principled case can be made for sequencing genes. Before running samples of genetic material through *gel electrophoresis*, researchers do change the structure of the gene: they break it up using enzymes, apply markers, etc. But they do *not* do this to see what happens to the organism if they break up their DNA. Rather, they do it so that they can read off the original genetic code once the electrophoresis is finished. This brings out another interesting feature of mere interactions: they are often used as *preparations* for other kinds of experiments (a point I already noted in chapter 12.2). Both staining cells and making particular genes are techniques that scientists use not only to examine the structure they are studying in more detail but also to find out where they should target subsequent interventions.[6]

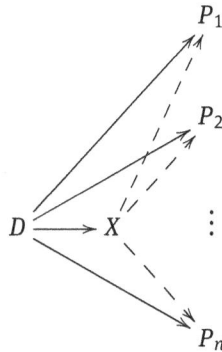

Fig. 14.1: A geometric property study. Manipulation D on X reveals geometrical features of X, possibly decomposes (dashed arrow) it into parts ($P_1, P_2, ..., P_n$) and reveals something about their geometrical features.

To summarize, in geometric property studies we *merely interact* with the architecture or geometry of a system giving rise to a phenomenon. We do not interfere with it to see what the consequences of our manipulations are. We know what the consequences will be, that is why we employ the manipulation in question.[7] We are not testing for a difference making relationship of any kind. Instead, we are

---

6 I will say more on the relations between different manipulations as I discuss other types of experiments.

7 This is true at least not once the method in question has been developed. As long as we are trying to devise new tools and techniques for research, even if they will later be used to merely interact, applying these not yet well-understood tools will often take the form of intervention

interested in the geometric properties our manipulation reveals. This knowledge can be referred to in the construction of scientific explanations. It can be used, e.g., to target specific parts of a system or mechanism with interventions, disambiguate relevance relations (especially causal vs. constitutive relations), and guide or constrain other kinds of mere interaction experiments. I will illustrate the interplay of different experimental manipulations in due course.

14.3 Composition Studies

Sometimes only a mechanism's parts are known while how exactly they work together to produce the explanandum phenomenon remains unclear. This is the case, for instance, when decomposition strategies have revealed the system's components but it is impossible to interact with the system as a whole in such a way that we can see how it is organized without decomposing it. If this is the case, scientists can try to deduce how the system as a whole is organized by putting the parts back together and investigating which possible arrangement shows the closest resemblance to the original system. Such studies I shall call *composition studies*.

To illustrate this type of study, we might consider James Watson and Francis Crick's (1953) discovery of the double-helix structure of DNA molecules. It took a good 80 years from the discovery of DNA to the identification of its structure. This was precisely because although DNA could be extracted and decomposed in various ways, nobody could assess the actual structure of the intact molecule. For subjecting it to biochemical analysis always broke it apart. However, scientists knew that DNA is made up of nucleotides, which, in turn, are made up of a nitrogen-containing base and a five-carbon sugar (together forming a nucleoside) and a phosphate group. A phosphate group can either be a purine (adenine or guanine) or pyrimidine (cytosine or thymine) where the total number of purines equals that of pyrimidines in any given string of DNA (this principle is known as Chargaff's rule). Supplementing this information with knowledge about molecular distances and bond angles, Watson and Crick eventually constructed their famous double-helix model:

> Using cardboard cutouts representing the individual chemical components of the four bases and other nucleotide subunits, Watson and Crick shifted molecules around on their desktops, as though putting together a puzzle. They were misled for a while by an erroneous

---

studies. For then we employ a new will-be tool to examine what its effects will be. But this makes for another discussion.

understanding of how the different elements in thymine and guanine (specifically, the carbon, nitrogen, hydrogen, and oxygen rings) were configured. Only upon the suggestion of American scientist Jerry Donohue did Watson decide to make new cardboard cutouts of the two bases, to see if perhaps a different atomic configuration would make a difference. It did. Not only did the complementary bases now fit together perfectly (i.e., A with T and C with G), with each pair held together by hydrogen bonds, but the structure also reflected Chargaff's rule [...]. (Pray 2008, p. 4)

The case nicely demonstrates that composition studies employ a kind of *reverse engineering*. In a way, they are the converse to geometric property studies: rather than looking for the entities and their arrangement (or organization), scientists know what the entities are and try to figure out what their arrangement should be so that the system as a whole fits together and exhibits the explanandum phenomenon.[8] figure 14.2 captures this. Notably, the manipulations of the components are not Woodward-style interventions: Watson and Crick shifted around the cardboard pieces; but once they knew the right components, they did not need to change anything about them to make them fit together. Note that the story I told here is not exactly historically accurate. Indeed, Watson already hypothesized that DNA was arranged in a double helix once he had seen x-ray diffraction images taken by Rosalind Franklin in May 1952. However, this does not defeat my point: Watson and Crick still fiddled with known components to figure out how to make them form a stable molecule. The suspicion that the structure they were looking for could well be a double helix may have guided their search; but it did not, all by itself, tell them how the molecules had to be arranged. This illustrates another important point, viz. that it is sometimes advantageous to combine different mere interactions. Taking x-ray diffraction images qualifies as a mere interaction along the lines of geometric property studies. If such a mere interaction guided Watson and Crick's composition study, the principle is likely to transfer to other cases. And indeed, we will typically know what the whole looks like and what the parts are before we try putting the pieces together. The aim of a composition study is to discover *how* to arrange these known parts such that they will form the target structure or exhibit the target phenomenon and share its properties. To find out about the components and to identify the target phenomenon and its properties scientists will usually employ other mere interactions such as time course observations and geometric property studies (see section 14.4).

More contemporary examples of composition studies can be found in *artificial intelligence* and *cognitive modeling*. Here too, scientists attempt to figure out how

---

8 For Watson and Crick, it was sufficient to show how their proposed configuration of DNA builds a stable molecule.

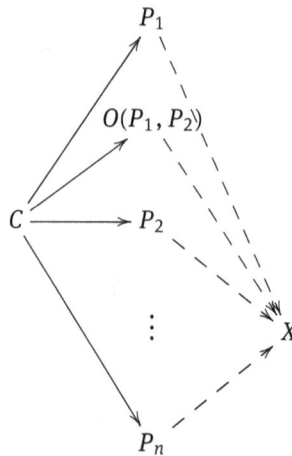

Fig. 14.2: In a composition study, scientists manipulate (straight arrows) how known component parts (P_1, ..., P_n) are arranged and how they interact to fit together and produce the phenomenon (X) under investigation (dashed arrows). In this figure, there is only one manipulation, C, representing the "putting together" of the component parts. In practice, there might by many different such manipulations each affecting a different subset of variables. Organizational features (arrangement and interactions) are represented by $O(P_m, P_n)$.

known functional units must work together to produce the phenomena they are trying to explain. Note that some modeling approaches start with analyzing a function and building a system that can accomplish that function quite independently of what the components are. Such "top-down" models (cf. Zednik & Jäkel 2014) are not quite what I have in mind when talking about composition studies.[9] Composition studies start with known components and aim to figure out how to arrange these components, and perhaps which of them are actually relevant to produce the explanandum phenomenon.[10] A case illustrating this latter use of composition studies is Izquierdo and Beer's (2013) model of C. elegans *klinotaxis*, viz. the spatial orientation behavior (particularly motion towards a food source) in a transparent roundworm. Izquierdo and Beer start developing their model by taking the known full connectome of C. elegans' nervous system and building a computational model of it. Using various optimization techniques and pruning

---

9 The notion "top-down" here is not the one I use when talking about interlevel experiments; see footnote 15 in chapter 2.3.

10 In a way, studies intervening into the organization of components (I discussed an example of this in chapter 8.4) are the interventionist analogue of non-interventionist composition studies.

algorithms, they cut down the number of nodes and edges in their model in such a way that it still exhibits worm-like klinotaxis behavior. The result is a model of C. elegans klinotaxis that is based on known components. It includes pretty much all and only those components (and, of course, their organization) relevant to the behavior Izquierdo and Beer sought to explain.

Both these examples demonstrate quite clearly how composition studies can shed light on something that gets usually lost if we use interventions alone, viz. the structure or organization into which components are embedded. To understand how a phenomenon is produced knowledge about the arrangement of components (and which components and organizational features are relevant) is vital. Note, though, that the point is not to use models and modeling approaches to get at these features. Scientists already know them before they enter into a composition study. Also, I am not saying that all modeling studies are non-Woodwarian in character. In fact, I have discussed a study by Lambon Ralph et al. (2007) in chapter 2.3.2.1 where a model network was systematically manipulated in different ways to elicit changes in categorization performance. This is clearly an interventionist study, albeit a modeling experiment. Likewise, we will often see interventions and interactions go hand in hand in building and testing models.[11] Yet, the nature of a composition study *per se* is clearly a non-interventionist one.

14.4 Time Course Observations

There are many ways to interact with a system or phenomenon in a non-interventionist fashion. Both geometric property studies and composition studies primarily serve to reveal structural information. However, mere interactions are not limited to this. We may, for instance, be interested in how a given system or phenomenon develops or unfolds over time. In this case, we may simply acquire data over time

---

11 There is an interesting contrast between intervention and composition studies. With interventions scientists often try to see what they can destroy before the system fails to exhibit the phenomenon under investigation. When the system fails to exhibit the phenomenon properly, the manipulated component is thought to be necessary for the investigated phenomenon. By contrast, composition studies are often designed such that components are added *until* the system exhibits the phenomenon under investigation. Therefore, composition studies may be viewed as a converse to intervention studies; they may serve as a kind of sufficiency test while intervention studies usually seem to test for the necessity of a component. If this is correct, there is a straightforward sense in which intervention studies will benefit from systematic combination with composition studies and vice versa; for, in the end, we want our explanations to cite all and only those features of a system that are necessary and sufficient for it to exhibit the phenomenon we are trying to explain.

without manipulating the system/phenomenon in any way other than what is required for the measurement. The measurement is, of course, assumed not to alter the very system or phenomenon we are interested in. It is a mere interaction, not an intervention. Studies that acquire data in such a way I shall call *time course observation studies*.[12] The crucial difference between time course observations and geometric property studies is what we observe: it is the sequence of events or temporal stages of an event, the dynamics of a process/phenomenon rather than the arrangement of physical parts.

As an example we can consider almost any behavioral study; be it on learning, monkey calls, or language acquisition. Likewise, we may think of Hans Berger's (1929) discovery of the *alpha-rhythm*. Planting electrodes on participants' scalps, Berger was able to record electrical potentials originating from their brains and found that, during rest, they exhibit a steady oscillatory rhythm at a frequency of 8–12 Hz—the alpha-rhythm. What is exciting about his discovery is not the fact that he was able to record electrical potentials (although he did invent the EEG as well), but that he measured an oscillatory pattern in a specific frequency range. Again, it is not the case that Berger's putting the EEG on a participant's head changed anything about the oscillations; the EEG merely made them visible. Thus, simple EEG recordings during rest do not qualify as a Woodward-style interventions; they are merely an interaction. As more contemporary examples we might consider the EEG studies discussed in chapter 2.3.1.4. The hypothesis that information learned during wakefulness is continuously replayed during sleep (cf. Buzsaki & Draguhn 2004) is mostly fueled by observations of oscillatory patterns over time. Similarly, functionally specific cortical oscillations have been identified mostly by observing these oscillations while participants engaged in specific tasks (e.g. Axmacher et al. 2006; Thut & Miniussi 2009). These, again, are instances of time course observation studies rather than intervention studies.

To further illustrate time course observations we may consider developmental studies or examples from genetics. Just as we can monitor neural signals, we can also monitor gene expression. Think about cases where gene expression is studied using *tracing techniques*, for instance. In such studies, markers binding to specific RNA sequences are introduced into the organism. If the relevant RNA sequences are present, the corresponding DNA must have been expressed in the nucleus, indicating that the relevant genetic sequence is active. Now once the markers have bound to the RNA, they can be made visible. What exactly needs to be done to visualize them depends on the kind of tracer being used. A fluorescent tracer, for in-

---

12 Another plausible name would be *monitoring studies* as a system/phenomenon is monitored over a certain period of time to observe how its behavior unfolds or capacities develop.

stance, may be visualized through exposure to ultraviolet light. There are multiple, and indeed highly complex, manipulations applied to the experimental organism throughout such an experiment. But for current purposes, it will be enough to just focus on the two most obvious steps, neglecting the overall complexity of the experimental setup: (i) the marker is introduced so it can bind to the RNA and (ii) the sample is manipulated such that the marker becomes visible to the human eye (say, it's exposed to ultraviolet light). What is important to note here is that neither of these manipulations interferes with the gene expression as such. The gene will be expressed no matter if we mark it or not and no matter if the marker is made visible or not. The use of these manipulations is that they reveal something we could not otherwise observe. They are not manipulations with respect to the gene's expression and thus do not qualify as Woodwardian interventions. Tracing genes is a non-interventionist manipulation. Still, monitoring when and how a gene is expressed clearly reveals information about causal processes within the organism.

The point becomes fairly obvious once we try to put an experiment looking at gene expression into interventionist terms. Suppose, for the sake of the example, that we are interested in seasonal FOXP2 (a gene known to be relevant to vocal communication) expression in songbirds. We get a bunch of finches and assess the expression of FOXP2 in different seasons using fluorescent markers. Now, what are the variables? Let S represent the season (taking values 1-4 over the course of a year), F the level of FOXP2 activity (taking values, say, between 1 and 10 depending on how much FOXP2 is expressed), M the introduction of the marker (1 for true, 0 for false), and L the light (0 for normal, 1 for ultraviolet). The relation we are interested in when doing this experiment is that between S (the seasons) and F (FOXP2 expression). But what we actively manipulate are L and M, none of which feature into the relation at stake. Manipulating L and M merely is a means of interacting with F in such a way to be able to assess its value. This is not relevant to the actual value of F and thus should not be considered an intervention. Yet, manipulating L and M is indispensable to the study of the relation between S and F (see figure 14.3).

Time course observation studies share an important feature with geometric property studies: both are essentially looking at the system or phenomenon of interest without requiring any extensive data analysis. It will hardly be possible to give the manipulations see in these studies an interventionist interpretation. Despite these similarities, they differ in an important respect: time course observations look at the whole phenomenon or system over time without taking it apart; geometric property studies are interested in the very composition or organization, not in behavior over extended periods of time.

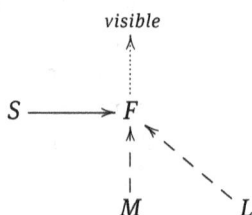

Fig. 14.3: Studying seasonal gene expression ($S \rightarrow F$) in finches by mere interaction ($M \rightarrow F$ and $L \rightarrow F$). Dashed arrows from M and L into F indicate that F is not altered by M or L, respectively but merely made accessible (visible).

Before moving on to the next type of experiment, I shall illustrate how time course observations can be combined with interventions.

14.4.1 Studying Plasticity

Just as we can study behavior and development by monitoring, we can also study how a system copes with different challenges. It may have to accommodate to a new environment or co-evolving predators, compensate for damage (e.g. after traumatic brain injury) or acquire new capacities (say learn a new language) to accomplish a given task. I already discussed examples involving neural plasticity and gene knockouts in chapter 8.4. Similarly, we may think about simple *conditioning* paradigms in this context. As an animal learns to associate a conditioned stimulus with an unconditioned one that causes an automatic response, it will eventually produce this response when faced just with the conditioned stimulus.[13]

Note that although these are cases of time course observation there is an important difference to the examples discussed above: there is an initial manipulation (e.g. acquisition of a brain lesion) that triggers or is otherwise relevant to the process we are monitoring. In this sense, studying neural plasticity is probably best conceived of as mixed mere-interaction and intervention study. For first there is a manipulation of the neural structure (say medial temporal lobe damage due to an accident) affecting a certain phenomenon (say, recognition memory). Then we observe how the system reorganizes and copes with the limitations incurred by the intervention. This second part then is a case of mere interaction while the first is an intervention into the neural structure with respect to a phenomenon. The same

---

13 This is basically what Pavlov famously did with his dogs (see Pavlov 1960 [1927]).

case can be made for gene knockout studies. Initially, a gene is knocked out in the developing organism. Subsequently, the organism's development is monitored and perhaps compared to healthy organisms, etc.

This illustrates how both interventions and mere interactions go hand in hand in empirical practice. Given the considerations thus far, we might get the impression that interventions usually illuminate our knowledge about components while mere interactions seem to be more apt to studying organizational features and temporal dynamics. However, this is not necessarily so; we can use both kinds of manipulations to learn about components and organization or temporal dynamics, respectively. For instance, as I argued in chapter 8.4, there are intervention studies designed to examine organizational features (recall, e.g., the study by Lambon Ralph et al. (2007) discussed in this context). Likewise, as the discussions of geometric property studies have made quite evident, we can study components by mere interaction.

14.4.2 Effects on Different Timescales

Another case where Woodwardian interventions and mere interactions in the form of time course observations get employed together is when we are studying effects at different timescales. Consider a patient suffering chronic pain whose plain levels vary (P taking different values between 1 and 9) over time. Two experimental drug treatments are possible. If effective, the first drug ($D = 1$) will develop its effect within the first few days after administration but it will wear off about one week later. The second drug ($D = 2$) works differently. It is administered on a daily basis. If it is effective, it will show improvements within a matter of hours that will last only for a day or two. Now administration of either of these drugs will count as an intervention on drug administration (D) with respect to pain level (P). However, it depends on the precise circumstances which of these interventions will be illuminating. And this is where interventions alone might be insufficient.

As said, the patient's pain levels naturally fluctuate without treatment (i.e. at $D = 0$). Suppose pain levels change slowly over the course of a few days but the pattern is not too regular. Now suppose we administer the first drug in this case and wonder if it was effective. The patient reports an improvement after, say, four days and an increase in pain after another six days. Now, these changes in the effect variable P might be attributed to the intervention (i.e. taken to be the effect of administering the first drug). However, they might equally well be due to the natural fluctuations we see in the patient's pain levels anyhow. Thus we cannot know if the drug was effective or not. Usually, scientists would disambiguate which of these conclusions is right by using a control condition. But these are not necessarily available in empirical, especially clinical, practice where single case studies are part an parcel of scientific research. If no control is available, the intervention just described was inconclusive: we cannot know (at least not based on observing drug-intake and pain levels) whether or not the drug was actually effective since we cannot disambiguate between drug-induced and natural fluctuations.

However, suppose the patient's pain levels naturally fluctuate on a daily basis. Say her pain is always least in the morning and worst at night. In this case, administering the second drug would lead to exactly the same problem as administering the first drug did in the scenario above: we cannot disambiguate between drug-induced and natural fluctuations. However, if we administer the first drug in this second case, we might get an informative experimental result; for now the natural and drug-induced fluctuations operate at different timescales. Natural fluctuations occur throughout the day with constant pain levels across different days for specified times of the day while drug-induced fluctuations occur over a period of multiple days. Thus, if our patient does report less pain in a few afternoons throughout the trial period than she usually does, we might consider this an effect of the drug rather than a natural fluctuation. Similarly, we can administer the second drug in the first case. While natural fluctuations occur over days, drug-induced fluctuations will occur (if at all) within hours. That is, if there is repeated short-term improvement in this scenario, we can attribute this effect to drug administration rather than consider it part of the natural fluctuations.[14]

What do we learn from this? First, we learn that interventions, even if effective, are not necessarily illuminating all by themselves. The observed effects may also depend on other, even completely independent causal processes. Ideal intervention studies would, of course, control for such confounding factors. But sometimes they are neither known nor is it possible to control for them. Second, we can compensate for such unilluminating interventions by cleverly combining interventions with time course observations. In the case I described, the clue was to know how the phenomenon (here: the pain level) develops over time and how the drug effects unfold over time. Once both are known, we can decide which drug to administer and monitor how the pain levels develop during the trial period. We can picture this as combining waves of different lengths (see figure 14.4). If the drugs operate on the same timescale as the natural fluctuations in the patient's pain levels and if the drugs are ineffective, the curves will remain unaffected in both cases (green and blue curves). But if the treatment is effective, and we pick the right combination (the second drug in the first scenario, the first drug in the second scenario), we will

---

14 Obviously, all of this assumes that the natural fluctuations are somewhat reliable and do not change midway through the clinical trial.

get a different curve reflecting the combined development of the phenomenon and intervention over time (orange curve).

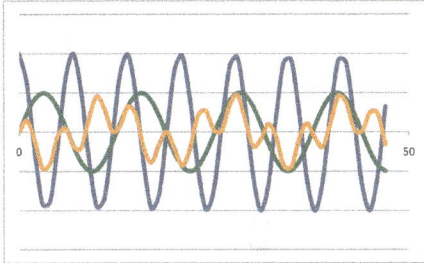

Fig. 14.4: A slow wave with low amplitude (green), a fast wave with high amplitude (blue), and a wave resulting from the conjunction of both (orange); time is shown on the x-axis. Green and blue waves correspond to natural fluctuations in pain levels in our example. The orange curve corresponds to the combination of either of these with a pain killer operating on the other timescale.

The interventionist might object that rather than requiring time course observation in addition to interventions we could, in principle, define our variable P such that it captures how pain levels develop over time. But this would require us to capture complex dynamic processes in a single variable. Such strategies are precisely what I criticize: we should not be tempted to squeeze and smooth all the experiments we encounter into an interventionist framework; we should consider their designs and employed methodologies in great detail to make sure we do not miss out on anything important. In the current case, the monitoring is an important feature of the experimental design that sets it apart from a simple intervention study.

14.5 Mining for Correlations

The studies considered so far did not require an extensive amount of statistical analysis; in fact, they were more like some form of enhanced form of passive observation from which conclusions could be drawn more or less immediately. The mere interaction studies I will discuss in this section are different: they require extensive data processing and transformation.

In some studies, scientists measure the values of a whole range of different variables within a single experimental setup and subsequently mine the resulting dataset for correlations. I shall call these *purely correlational studies*. From the correlations they identify, scientists can construct mathematical or computational models using different algorithms—typically linear or network models—that can be used for predictive purposes or the identification of functional units, respectively, independently of whether or not the variables map onto something "real".

The following simple example illustrates how purely correlational studies work. Suppose there is a group of subjects (S_1, S_2, ..., S_m). Each subject performs a number of cognitive tasks (T_a, T_b, ..., T_n) while their performance is monitored. Plugging the performance data into a computer, we can calculate how strongly performance on one task correlates with performance on another. The results can be visualized in a correlation matrix (see figure 14.5). Note that while some manipulations may be necessary to acquire the data that is subsequently mined for correlations, these are not Woodward-style interventions; they are mere interactions just as in time course observations and geometric property studies.[15] Unlike these other mere interaction studies, however, correlation mining requires extensive data analysis before any conclusions can be drawn. But this analysis does not qualify as Woodward-style intervention either; for it does not change anything about the data. It is best conceived of as a (mathematical) search for similarities and dependencies, or, more precisely, correlations.

The very same principle can be used for variables representing any kind of information: socio-economic factors, data on medical care and prevalence of illnesses, demographics, climate data, or brain activations in different voxels. In what follows I shall illustrate purely correlational studies by looking at two different examples.

14.5.1 Estimating Predictive Models

Using sophisticated mathematical tools, scientists can construct predictive models from purely correlational data. Recall the example of the Sally-Anne-task outlined in chapter 9.1. To assess children's' mentalizing abilities, scientists show them a little play with two puppets and ask them (the children) about the puppets' beliefs. In addition to performance on this *false-belief task*, scientists measure a whole bunch of other variables such as age, daily play time, general intelligence, language ability, number of siblings, and so on (see e.g. Sodian 2005; Poulin-Dubois et al. 2007). All of these (possibly correlated) variables are fed into a mathematical analysis to find out which of the observed variables can be used as predictors for the occurrence of the explanandum phenomenon (in this case children's passing

---

15 In some cases, these mere interactions may be given an interventionist interpretation and thus qualify as pseudo-interventions. I will discuss this in more detail in the sections 14.6 and 14.8 below.

$$S_1: \qquad \{T_a, T_b, \ldots, T_n\}$$
$$S_2: \qquad \{T_a, T_b, \ldots, T_n\}$$
$$\vdots \qquad\qquad \vdots$$
$$S_m: \qquad \{T_a, T_b, \ldots, T_n\}$$

$$Corr = \begin{pmatrix} 1 & corr(a, b) & \ldots & corr(a, n) \\ corr(b, a) & 1 & \ldots & corr(b, n) \\ \vdots & \vdots & \ddots & \vdots \\ corr(n, a) & corr(n, b) & \ldots & 1 \end{pmatrix}$$

Fig. 14.5: In purely correlational studies, correlation matrixes are calculated. $corr(a, b)$ is the correlation coefficient for performance on tasks T_a and T_b (say, averaged over all subjects S_1, S_2, ..., S_m). The diagonal of the matrix is 1 because each data point is perfectly correlated with itself.

of the false-belief task). An example for such a mathematical analysis is a *principal component analysis* (PCA). This method involves finding a linear combination of (a subset of) the analyzed variables that has maximum variance. It goes through multiple iterations and will eventually transform the set of (possibly correlated) observed variables into a set of linearly independent variables. Using these, we can generate equations relating analyzed factors to the outcome we want to predict.[16]

Put crudely, this is how intelligence tests work, too. Scientists gather data from a population of individuals and assess their performance on various cognitive tasks (these are the T-variables) as well as their general intelligence. Statistical analysis then allows them to find performance on which tasks serves as a good predictor of general intelligence and thus can be used for its assessment in an intelligence test.

Similarly, risk factors for, say, suffering Alzheimer's disease can be identified using this method. Just take a group of people, assess, say, how much physical activity they engage in, how much TV they watch, how much they read, what their level of education is, whether their relatives suffer from dementia, whether they smoke, eat a lot of meat, go to church regularly, etc. Then see who suffers from Alzheimer's disease and to what degree and feed that information into your

---

16 The precise mathematical details are not important here, but see e.g. Field (2009).

statistical analysis. If everything works out nicely, the analysis will return which (set) of the assessed factors makes for a good predictor of Alzheimer's disease.

14.5.2 Identifying Functional Units

A similar, albeit different way to "mine" correlational data to reveal dependency relations is to use them to build networks. In these networks, factors (i.e. measured variables) are represented by nodes and their correlations by edges of varying strengths. The strength of the edges corresponds to the correlation coefficient. We may further cluster strongly correlated factors together and infer that these clusters represent functional units.

A case in point is research on resting state functional connectivity (Power et al. 2011). Acquiring resting state fMRI data clearly does not change anything about the connectivity of the brain; it is a mere interaction rather than an intervention. Once the data has been acquired, we can treat the recorded BOLD signal[17] as an indicator of the neural activity in each voxel. These are the variables, the values of which we can subject to a correlation analysis. The correlation matrix will thus have as many rows and columns as we have voxels. Taking the values from the correlation matrix, voxels can be plotted as nodes and correlation coefficients as edges of different strengths to construct a network model of the brain. Subsequent application of clustering algorithms puts those voxels that are more concurrently active closer together and pushes those showing less correlated activity further away from one another. The resulting clusters may now be interpreted as functional units, as components of the functional machinery the brain provides.

Note that quite unlike geometric property studies this procedure does not have much to do with anatomy. Voxels are essentially chunks in space which happen to be filled by portions of the brain we stick into the fMRI scanner; the brain is cut up into chunks without respecting anatomical features.[18] These are very different ways of cutting up the brain, based on different criteria.

The resulting computational models thus are relatively theory-free; at least they are insofar as the selection of variables was to begin with. They are independent of any physical realization of the variables in the real world and completely independent of whether or not these map onto meaningful structures or actual causal factors. Yet, these models are predictive and they do give us valuable infor-

---

17 The blood-oxygen-level-dependent signal is used as an indirect measure of brain activation in functional imaging.
18 Yet, at least in the case of Power et al. (2011), the resulting network shows a remarkable resemblance to anatomically and functionally defined brain area networks.

mation to contribute to our construction of scientific explanations: they help us identify which factors are potentially relevant and help generate (causal) research hypotheses. And, again, this they do without any Woodward-style interventions. All of this is just mining data that has been acquired by mere interactions.

14.6 Simulating Interventions: Difference Association & Dissociation

As mentioned above, mere interactions and Woodward-style interventions often go hand in hand in empirical practice. But there is another interesting way in which interventions and mere interactions can be related. As mentioned in chapter 12, researchers can sometimes interpret mere interactions in a Woodwardian fashion. Such *pseudo-interventions* may be considered as a separate kind of manipulation as they are different from both simple (primarily observational) mere interactions and genuine interventions. However, it is important to recognize that the individual manipulations on which pseudo-intervention studies are based are simply mere interactions: scientists measure, observe, correlate, etc. They do often manipulate in order to acquire the data they are interested in, but they do not do this by inter-ventionist wiggling-experiments. Pseudo-interventions are best understood as *a way of interpreting sets of mere interactions*. In this sense, pseudo-interventions are reducible to mere interactions. Figure 14.6 graphically depicts the relation between different modes of scientific inquiry. Typical pseudo-intervention studies

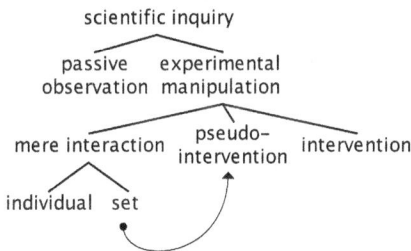

Fig. 14.6: Modes of scientific inquiry. Passive observations contrast with manipulations which can be mere interactions, pseudo-interventions, and interventions. The fact that pseudo-interventions are in fact sets of mere interactions interpreted as if there had been a genuine intervention is captured by the arrow moving sets of mere interactions directly below manipula-tions.

simulate interventions by comparing and contrasting, e.g. across experimental groups or conditions. Such simulated interventions are particularly useful in cases where direct interventions are not available. Say, for instance, we want to assess differences in cortical language processing between native speakers of different languages (see chapters 2.3.2 and 9). There is no experimental intervention changing a native speaker of one language into a native speaker of another. So to assess the differences in cortical language processing between them, scientists resort to *comparisons*. Systematic comparisons (contrasts) can be drawn both between groups or experimental conditions. Depending on the precise experimental setup, studies varying experimental conditions might still be considered as genuine intervention studies in some cases. However, as soon as the desired differences in the putative cause-variable cannot be produced by intervention, scientists rely on pseudo-intervention studies instead. *Difference association studies* are a case in point; they associate difference observed in one variable with difference observed in another. To do so, they do not need interventions but associate the results of different comparisons with one another instead. The data on which these comparisons are based is usually obtained through mere interactions. It is then evaluated and interpreted *as if* there had been Woodwardian interventions.[19]

For illustration consider researching the effect of medial temporal lobe (MTL) lesions on spatial memory. Rather than testing participants on spatial memory, subjecting them to MTL lesions and then re-assessing their memory skills, scientists can compare the spatial memory abilities in patients with MTL lesions to that of healthy controls.[20] Figure 14.7 (a) depicts this graphically. Now the question is whether we should think of this kind of study as a Woodward-style intervention or a mere interaction experiment. On the one hand, the data acquisition is intervention-free just as is the data acquisition in the experiments described when discussing geometric property studies and correlation mining. Although there is a lot of interaction involved in assessing spatial memory skills and taking structural brain images to localize patient's brain damage, etc. this kind of data acquisition as such is not an intervention in Woodward's sense. Rather, what happens is that scientists associate the difference identified in one domain (MTL lesion vs. no lesion) with differences found in another (spatial memory deficit vs. unaffected performance) in a way similar to what happens when we mine for correlations.

On the other hand, it is important to recognize that in difference association studies of this kind the data is often presented and interpreted *as if the difference in one domain resulted from an intervention into the other*. This might even be

---

19 The trick to make difference association look like an intervention rather than a mere interaction is to choose the variable set differently.
20 This kind of comparison follows essentially the *subtraction logic* introduced in chapter 7.

$$
\begin{array}{ccc}
P & H & [H - P] \\
C_P & C_H & [C_H - C_P] \\
\vdots & \vdots & \vdots \\
& & \\
B_P & B_H & [B_H - B_P]
\end{array}
$$

$$I \longrightarrow B \longrightarrow C$$

$$I \longrightarrow C \longrightarrow B$$

Fig. 14.7: A difference association study. Top: Comparing status of brain structures (*B*) with cognitive performance (*C*) across individuals (patient *P*, healthy control *H*). Bottom: Interpretation as an intervention study.

the most salient difference between simple correlation mining and difference association: the very fact that the comparisons we draw are used to *pretend* genuine Woodward-style interventions. This is why I call the manipulations we see in difference association studies pseudo-interventions. Experimenters do not simply collect data and feed them into an automated analysis but they evaluate and contrast acquired data is if there had been genuine interventions.[21]

At this point, a few remarks are in order. First, difference association and pseudo-interventions are not specific to fMRI. Recall the discussion of mere interactions using EEG in section 14.4 above. As described, we can use EEG to observe neural oscillations over time. This is a mere interaction. However, depending on the overall experimental design scientists employ, we may also think of EEG studies as pseudo-interventional. If we compare EEGs taken in different conditions or subjects, for instance, we may enter into a pseudo-interventionist context, or— depending on the experimental design—even a genuine intervention context. For illustration consider an epilepsy patient. Suppose we are interested in how the EEG signal changes during a seizure. To investigate this, we record EEG signals in the patient while she is not suffering a seizure, then induce a seizure and continue measuring EEG signals. This would be an intervention study from which we learn that changing from non-seizure to seizure situation made the EEG pattern change

---

21 Note though, that once obtained data can be analyzed in multiple different ways. I will say more on this in due course.

in such and such a way. Similarly, if we are studying how resting state EEG is different between our patient and a healthy control, we may tell an interventionist story: changing from healthy control participant to epilepsy patient results in characteristic changes in the resting state EEG. Though this sounds like an intervention study, we did not actually intervene. We merely interacted as we measured the EEG in both participants and *simulated* an intervention in our analysis by contrasting the data from patient and control. Therefore, this qualifies as a pseudo-intervention study, viz. a study that can be given an interventionist reading despite the fact that there are no Woodwardian interventions. Note that both these scenarios are importantly different from simply recording a rhythm: in both these cases we are interested in the changes in the EEG *as a result of our (pseudo-)intervention.* Pure mere interaction studies are different in this respect: they are not interested in a specific contrast but in observation of a given state, system, development, etc. Time course observations studies, geometric property studies, etc. are studies that are not concerned with the effect of a (simulated) manipulation but merely use manipulations as observational tools.[22]

Second, whether a mere interaction is interpreted as a pseudo-intervention may depend on the precise experimental design we employ. It is not a question of doing something different but of asking a different question and examining acquired data in a different way. In fact, a data set acquired in a pseudo-interventionist study may be used for correlation mining later on. Or different contrasts may be calculated from the same dataset depending on the research question we are asking (see also section 14.8). Say, for instance, we are measuring a number of variables (gender, body size, weight, age, intelligence, socio-economic status, ethnicity, physical activity, general health, diet) in a group of people over time. If we contrast, say body sizes of males and females, we may interpret that as a pseudo-intervention into gender with respect to body size. However, we may also perform a statistical analysis (as described in section 14.5.1) to generate a mathematical model that allows us to predict body size from the other variables we measured. Either way, the measurements we take are mere interactions. We do not change anything about our sample population, we merely interact with it. Subsequently, we can analyze acquired data in different ways. Some of these analyses will suggest interpreting our mere interactions as if there had been genuine interventions (viz. as pseudo-interventions) while others will not. Similarly, almost any experimental technique can, at least in principle, be used for interventions, pseudo-interventions, or mere

---

22 Obviously, different observational tools may incur different artifacts and have different limitations. But as I said earlier, this is not the place to discuss them. Besides, the precise practical limitations have no bearing on the principle point I am making here.

interactions alike depending on the given research context and the research question scientists are trying to answer. I will say more on this below (in section 14.7). Like interventions and simple mere interactions, pseudo-interventions can occur isolated, or in combination with different manipulations.

Third, note that the interpretation of mere interactions as interventions raises a peculiar problem: observed correlations and associated differences do not give us directionality. If we want to interpret differences in one domain as an effect of (hypothetical or pretended) interventions into another, we must postulate which of the two domain depends (causally or otherwise) on the other. Where we carry out an actual intervention, we know something about the direction in which the (causal) dependence relation supposedly holds: if a difference in B brings about the difference in C, we can manipulate C by intervening into B but not (or at least not necessarily) vice versa—and, of course, the same holds true the other way around. Correlations alone, by contrast, cannot tell us anything about the direction of the (causal) dependence relation (see figure 14.7).[23]

In our example, perhaps for all of cognitive neuroscience, we have a straightforward strategy to handle this problem: the dominant conception is that brain processes implement cognitive processes and we know that brain lesions quite frequently affect cognitive processes. Given this, and perhaps also owing to a tendency to not like mental-to-physical causation, we will speak of the brain lesion as bringing about the cognitive deficit.[24] Yet, by associating the differences alone, we could never conclude this. It is solely our intuition (plus perhaps our neurophysiological education or scientific conventions) that renders the interpretation of a cognitive deficit causing a brain lesion odd.

Despite this weakness, the very fact that difference association studies do not require genuine Woodward-style interventions is advantageous: it allows a broader range of phenomena to be studied as we can even investigate phenomena for which no intervention studies are available for practical, epistemological, or ethical reasons. The lesion studies I mentioned a few times now are illustrative: they are only available in humans where lesions occur naturally. While studying patients with cortical lesions can be very illuminating, there usually is complication: patients are typically only assessed thoroughly after they acquired their brain lesion but not beforehand. Thus, the assessment of their deficits draws heavily on comparisons to healthy individuals undergoing the same test. This is difference

---

23 I have discussed this issue in great detail in part II. See especially chapters 7 and 8.

24 If this "bringing about" relation is actually a causal one is highly debated. The perhaps least controversial assumption is that the mental *supervenes on* the physical (see also chapter 4.2).

association between patients and healthy controls rather than an intervention study in the patient we are considering.[25]

It is also worth noting in this context that difference association does not have to be limited to observing a single difference. In fact, we could observe a whole range of differences between two populations or a systematic pattern of differences between two patients suffering from different brain lesions. In fact, this is what Broca and Wernicke did when comparing the cognitive deficits and associated brain lesions in their patients to draw conclusions about the functional organization of the human brain (see chapter 2.2).[26] The whole business of reasoning about dissociations to delineate phenomena (see excursus on page 41) essentially builds on observations and difference associations.

14.7 Multiple (Pseudo-)Interventions

Many if not most difference association studies (particularly in cognitive neuroscience) assume that we can map differences in function more or less neatly onto differences in architecture. This, however, might not be the case as recent debates on *dynamical systems* make evident.[27] But even when we assume that it is in principle possible to decompose a complex system such as the brain into functional subsystems, the components our methodologies suggest might not respect functional units.

The case of neuroimaging with fMRI is illustrative. Although most studies using this method presume some kind of decomposability of the brain into functional networks, the brain is cut up into voxels—chunks of space—rather than neural populations. This holds true even for studies aiming to identify functional networks like the one by Power et al. (2011) discussed above. It is a notorious problem for fMRI studies due to the limited spatial resolution of imaging data (though it is still much better than that of many other neuroimaging techniques, see figure

---

25 For more on the limits of intervention studies see chapter 9.1.

26 Some may object that we should consider patients' lesions as "experiments of nature" and that as such they are interventional in character. I take it, though, that they are at best pseudo-interventions. These lesions have not been produced under controlled conditions to see what their effects will be and we only assessed their effects by indirect means, viz. by comparisons to healthy controls.

27 I discussed dynamical models briefly in chapter 3.7 as well as in chapter 13.4. There is, admittedly, much more to be said about the explanatory status of dynamical systems but this is not the place to do so. For current purposes just note that there are cases where we might not be able to decompose a system into functional units for whatever reason.

13.1 in chapter 13.3). But cognitive neuroscientists have found a way to overcome this limitation, at least partially. By using *adaptation paradigms* (e.g. Lingnau et al. 2009; Krekelberg et al. 2006) they can figure out whether the neurons within one voxel actually belong to two different populations that selectively respond to different stimuli.

The basic idea of behind adaptation is that neurons *get used* to a certain kind of stimulus if it is repeatedly presented. So if we keep presenting, say upwards moving dots to a participant, we will initially see high activity in visual motion areas responsible for processing upward motion but the response will reduce when upward moving dots are repeatedly shown. For downward moving dots, we will see no interesting effect in this area as long as it is really just responsible for processing upward motion. The same goes for downward moving dots and the neural activations measured in visual areas responsible for processing downward motion: we see adaptation for downward motion but no interesting effects for upward moving dots. But how about a voxel that contains neurons of two different populations responsible for processing downward and upward motion, respectively? If we show just one stimulus repeatedly, we will see adaptation just as if there was only one population. So this alone will not be conclusive. If we observe some adaptation for both upward and downward moving dots we might be lead to conclude that the neurons in the voxel we are looking at collectively respond to vertical motion more generally, irrespective of direction. What else could we do? Well, we can alternate the presentation of the stimuli. If we do this and find no adaptation this permits, again, two different conclusions: either the neurons in the voxel do not respond specifically to vertical motion at all, *or* there are two separate populations each responding differentially to either upward or downward motion. Because the stimuli are alternated in this setup, we will not see adaptation. This is because separate sets of neurons get alternately activated. By contrast, if the entire population were responsive to vertical motion more generally, we would also see adaptation in this case. Taking together the evidence gained in all three experimental setups (i.e. repeated upward motion, repeated downward motion, alternating upward and downward motion), we can now answer the question whether a voxel contains functionally different components (i.e. different neural populations): if there is no adaptation in when we alternate between stimuli but adaptation to each individual stimulus we can infer that there are indeed two functionally specialized subpopulations in the same voxel. This illustrates how a clever and systematic combination of experimental manipulations can be illuminating where individual manipulations are not.[28]

The designs scientists employ with functional imaging adaptation paradigms effectively rely on difference association: for each condition, responses associated with sequences of presented stimuli get compared. We might want to interpret this in a Woodwardian fashion and hold that changing between conditions is an intervention with respect to the BOLD signal measured for the voxel in question. Be that as it may, the important point is that the observations we make from any single experiment (repeated upward motion, repeated downward motion, alternating upward and downward motion) will not be conclusive. What it takes to disambiguate competing hypotheses is to look at all of these experiments *together*. This illustrates how multiple pseudo-interventions (and likewise proper interventions) can reveal e.g. organizational features of a system that are not accessible by using a single pseudo-intervention (or proper intervention) alone; perhaps not even accessible by a geometric property study. This is another feature of experimental practice that is often neglected when we think about how (mutual) manipulability leads us to explanatorily relevant information: manipulability of some factor Y through some factor X alone is not necessarily informative; it might take multiple manipulations to figure out what is going on—and they might have to be cleverly combined (see figure 14.8).

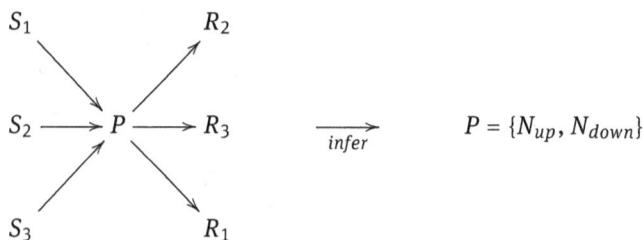

Fig. 14.8: Multiple different (pseudo-)interventions (different stimuli, S) applied to one spatial part (P) of a mechanism produce differential responses (R). Systematic comparison of the different responses associated with different stimulations leads researchers to conclude that P in fact contains functionally different components (different neural populations N_{up}, N_{down}).

---

28 This point is a general one. It holds for all kinds of different manipulations. Since adaptation paradigms rely on difference association, the case at hand is probably best characterized as a multiple pseudo-intervention design.

On a related note, the example I just described also illustrates that pseudo-interventions (and the same goes for proper interventions) can be used not only to uncover causal relations but also to develop hypotheses as to where further decomposition of a system might be needed to reach a satisfactory explanation of a given phenomenon.

14.8 Mixed Cases: Interaction Effects

The basic rationale behind difference association studies can be extended to contexts where multiple comparisons are needed. This is the case, for instance, when we want to assess differences in multiple domains and their interactions. To do this, scientists often calculate *statistical interaction effects*. This assessment of interactions between different factors relevant for a given phenomenon is of great importance for the explanation of more complex phenomena such as, e.g., language processing.

The "Sights & Signs" project sketched in chapter 2.3.2.3 is an example where lots of different statistical interaction effects can be assessed. But for simplicity's sake let's just work with a toy example here: an fMRI study that investigates the neural basis of *reading processes* in dyslexic and healthy patients. Dyslexia is a developmental condition; although it does not affect general intelligence, patients suffer from difficulty in learning to read or interpret words, letters, and other symbols. To examine how the neural processes implementing reading are affected by dyslexia, we first need to recruit two groups of participants (dyslexic patients and healthy controls) matched for relevant factors such as age, education, intelligence, and so on. Then we need all participants to perform two tasks while lying in the fMRI scanner (a reading task and a baseline task). This type of design is known as a 2×2 design. It can be visualized in a contingency table (see figure 14.9).

Taking the fMRI scans is simply a way of measuring the thing we are interested in, i.e. the brain activity. Thus, although there is a lot of interaction involved in taking the images, the data acquisition (taking the images) as such is not an

	dyslexic patient	healthy control
reading task	α	β
baseline	γ	δ

Fig. 14.9: Table showing a 2×2 design.

intervention.[29] When we analyze the data, what we do is calculate contrasts. This, too, is not in itself an intervention; for we do not change the data, we merely contrast different portions of it with one another. To assess the *main effects* of group and task, respectively, we assess what overall difference group and task make to the recorded brain activations. To assess the main effect of dyslexia, we contrast, across both tasks, the activations measured in patients with those measured in healthy controls. Likewise, to assess the main effect of reading, we contrast the activations measured during reading with those measured during the control task in both groups. So far, this is just difference association in a pseudo-interventionist fashion all over again. The imaging study may be given an interventionist interpretation even though the experiment itself was merely interaction-based.

But this toy example permits more than just looking for main effects. When we are interested in how the neural substrate of reading is affected by dyslexia, we are asking for an *interaction effect*, viz. the interaction of dyslexia with brain activations during reading.[30] How do we assess this? It is a matter of multiple comparisons. First, we need to know what the main effect of reading is in each group. To assess that, we calculate the difference in brain activations between reading and baseline tasks separately for each group. Second, we want to know how this main effect is different in our two groups. In order to assess that, we compare the results of the first analysis between both groups (see figure 14.10).

$(\alpha + \beta) - (\gamma + \delta)$	main effect of task (reading)
$(\alpha + \gamma) - (\beta + \delta)$	main effect of group (dyslexia)
$\alpha - \gamma$	effect of reading in dyslexic group
$\beta - \delta$	effect of reading in control group
$(\alpha - \gamma) - (\beta - \delta)$	interaction of group & task

Fig. 14.10: Contrasts being calculated based on the design shown in figure 14.9.

Although the kind of manipulation is essentially the same in both steps, this analysis is interestingly different from simple difference association described in 14.6. Given our research question, the pseudo-intervention of interest is the second contrast calculation. The first calculation (isolating reading-associated activity

---

29 As I will discuss below, we may think of administering different tasks as an intervention, though.

30 It is important to emphasize that "interaction effect" refers to the statistical effect of an interaction of separate factors and does *not* mean the effect of a manipulation that I consider a mere interaction. Also, there is much more to be said about the math behind interaction effects. For current purposes, this toy illustration based on fMRI contrast will do, though.

by comparison to baseline activity) is merely providing the basis for the group comparison to be performed. This changes the role of the first comparison. Taken individually, the first contrast assesses (in a difference associative manner) the effect of reading. It may thus be interpreted as an intervention into task, though separately for each group. However, when we are interested in the second comparison only, the first comparison comes to play the role that used to be played by the original data in the first comparison. All of a sudden, the first set of calculations is no longer a matter of simulating interventions. Rather, it becomes a preparatory step and we now interpret the second contrast calculation as an intervention. For it is this second pseudo-intervention that now marks the relevant difference with respect to the question of how dyslexia interacts with brain activity during reading. But what then is the role of the first contrast calculation? In a way, we may consider it analogous to setting background conditions or holding certain factors fixed. It is a preparation that enables us to carry out the desired manipulation—in this case, a pseudo-intervention illuminating how the cortical processing during reading differs between dyslexic patients and controls.

This example illustrates quite nicely that whether or not a given manipulation is interpretable as a Woodwardian intervention with respect to a given phenomenon depends on the specific research question we are trying to answer. This is not surprising; after all, it is a feature of Woodward's account that it supports contrastive explanations, viz. explanations that relate changes in some variable X (due to intervention I) to changes in another variable Y (see chapter 4). Accordingly, changes in different variables will have to be considered depending on which variables we ask about. In fact, this holds true not only for pseudo-interventions but also for genuine Woodwardian interventions: the very same experimental manipulation may count as an intervention with respect to some variable but not others. Therefore, the very same experimental manipulation may be an intervention with respect to some phenomenon in one study but serve as a perhaps preparatory mere interaction in another. A case in point are genetic manipulations in C. elegans.[31] Scientists can

---

31 Ken Waters (2014) discusses this case in the context of distinguishing between actual and potential difference makers. According to Waters, an *actual difference maker* is the cause variable that actually made the difference to a given effect in a given scenario. By contrast, a *potential difference maker* is a causally relevant variable that could, potentially, also have made the difference in question but did not do so in the scenario in question (see Waters 2007). In a way, the transition between mere interactions and pseudo-interventions might be analogous to Waters' distinction between potential and actual difference makers. For holding potential difference makers fixed (by merely interacting) ensures that we can identify the effect of actual difference makers. Yet, there is an important difference between his distinction and mine: Waters only considers those manipulations that are potential interventions if different questions are to be answered; the does not take into account such things as geometric property and purely observational studies.

systematically examine the effects that the synthesis of certain proteins has von axon growth by manipulating the genes encoding these enzymes. Experiments in which they knock out those genes, we may consider as classical intervention studies in Woodward's sense, perhaps combined with some time course observation. However, proteins are not the only thing that is causally relevant to axon growth in C. elegans. Scientists observed that those worms displaying more axon growth were also more active, viz. moved around more. Thus, if we want to be certain that the effect we see in knockout worms is due to the experimental modification of genes encoding proteins, we must make sure to *control for motion*. But how can we do this? Geneticists have a simple answer: disable genes relevant for motion so the worm will no longer move around. If we knock out the genes relevant for motion to see what the effects on C. elegans' motion behavior are, we count this gene knockout as a Woodwardian intervention. The case is perfectly analogous to knocking out protein genes. However, if we knock out motion genes to see what the effects of interventions into protein genes are, knocking out the motion genes is no longer a Woodwardian intervention with respect to the phenomenon to be explained; for we are now asking about the effect of the protein on axon growth, not about the effect of motion.

This case is analogous to calculating interaction effects in functional neuroimaging. The very same manipulation can be counted as a Woodwardian intervention (or at least a pseudo-intervention) in one experiment while it is merely an interaction—something preparatory, a kind means to set background conditions or fix certain variables—in another. Importantly, it is not the experimental manipulation that is different, it is the *use to which we put it* within in our overall experimental design.

14.9 Manipulation Checks

I already pointed out that manipulations can play different roles depending on the context in which they get used. Knocking out the very same gene or comparing the very same data may function as a (pseudo-)intervention or a (preparatory) mere interaction depending on what research question scientists are trying to answer and what the overall experimental design looks like. When calculating interaction effects, we are employing mere interactions to set baselines for comparisons or to ensure certain variables are held fixed. However, we can further employ mere interactions to verify the effects of experimental interventions, viz. use them as *manipulation checks*.

To illustrate the role of manipulation checks let us look at some research in psychology. Psychologists are often interested in how emotions, mood, or affect, influence specific cognitive processes.[32] For instance, Storbeck & Clore (2005) investigated whether *false memory effects* can be modulated by affect. False memory effects can be produced when participants are presented with a list of semantically related words that are all highly associated with a single non-presented word (cf. Roediger & McDermott 1995). Upon free recall, participants are likely to include this word in the list of items they "remember". An example would be the following list: *doze, snore, bed, blanket, drowsy, tired, nap, snooze, rest, dream*. In this case, the predicted false memory item (the "lure") would be *sleep*. Participants who learned *doze, snore, bed, blanket, drowsy, tired, nap, snooze, rest, dream* are likely to name "*sleep*" as an item on the list during recall although it has not been present at all.

Starting from the observation that a positive mood seems to facilitate memory encoding, Storbeck & Clore (2005) investigated how affect modulates the occurrence of false memory effects. In order to do so, they manipulated participants' mood by making them listen to either peaceful (positive affect) or aggressive (negative affect) music. After performing the experimental memory task during which general performance and false memories were assessed, the experimenters carried out a *manipulation check* to verify that participants were indeed in the intended mood: they administered a questionnaire assessing participants' mood. The basic structure of this setup is depicted in figure 14.11.

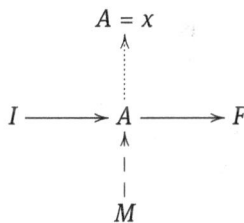

$$A = x$$

$$I \longrightarrow A \longrightarrow F$$

$$M$$

Fig. 14.11: The effect of affect (*A*) on false memory (*F*). A manipulation check *M* serves to measure the value of *A* and thereby ensures the validity of the intervention *I* on *A* with respect to *F*. While *I* → *A* is a Woodward-style intervention, *M* → *A* is a mere interaction.

As result of their study, Storbeck and Clore report that (i) their manipulation of mood was successful, and (ii) that participants in a negative mood "were significantly less likely to show false memory effects than those in positive moods"

---

32 In psychology, "affect" is conceived of as a pre-emotional disposition that either positively or negatively influences overall mood.

(Storbeck & Clore 2005, p. 785). There are several interesting features to note about this design in the context of distinguishing different kinds of manipulations. The major conclusion was that affect indeed modulates false memory effects, viz. (ii). This has been demonstrated through manipulation of affect and the observation that changes in affect indeed incur changes in false memory rates. This fits nicely with the interventionist picture: inducing affect is an intervention on mood that made a difference to false memory effects.

If we consider the exact experimental procedure, it becomes clear that to establish (ii), the experimenters had to establish that their manipulation worked in the intended way in the first place. This is why result (i), i.e. the result of the manipulation check, is vital. But result (i) has not been obtained through a Woodward-style intervention with respect to affect. Rather, what we see is a mere interaction: a questionnaire is administered and the results are evaluated to assess participants' mood, not to manipulate it or anything that depends on it. In this sense, it is like making the membranes of cells visible to ensure you can induce a current into them.[33] The questionnaire merely serves as a tool to measure the effect on participants' mood thus *enabling* the conclusion drawn in (ii).

Although the case I discuss here is not entirely intervention-free (there is a Woodward-style intervention after all), it nicely demonstrates the importance of mere interactions. It was only through the mere interaction that the intervention could be verified. Thus, it is only due to the non-interventionist portion of the design that the experimental strategy employed allows us to draw any conclusions at all from the applied interventions. Note that while the function of manipulation checks is similar to that of preparatory experimental manipulations such as setting certain variables or background conditions, there is an important difference: with manipulation checks, we merely measure or observe the value a certain variable has taken but do not interfere with it, i.e. we do not set it to a certain value. Yet, manipulation checks share an important feature with all the other kinds of mere interactions discussed thus far, be it preparatory manipulations fixing certain factors or such things as slicing, staining, and marking: they help us gain additional experimental evidence on the basis of which further (perhaps genuine Woodward-style oder pseudo-interventionist manipulations) can be carried out and evaluated. This highlights what has dawned on us since part II: mere interactions are—at least sometimes—vital for experimental practice, and they may lay the groundwork for carrying out genuine Woodward-style interventions.

---

33 In fact, this case brings out the concept of mere interaction even more strongly as the examples discussed thus far; for participants' mood is not even manipulated at all! It is measured, no more than that. The manipulation of participants' mood is something else, i.e. making them listen to different kinds of music.

14.10 Translational Paradigms

Before closing, I want to draw attention to a final type of experiment that might also be considered preparatory to genuine interventions. I already mentioned *(computational) models* and *model organisms* a few times. Especially where scientists meet practical or ethical limitations, research can benefit from transferring experimental paradigms and hypotheses from a target system (e.g. a human) to a model organism (e.g. a rat or monkey) or a (computational) model. Once this *translation* has been achieved, interventions can be carried out that are otherwise unavailable. In turn, the obtained results are transferred back to the target system to illuminate the original research question. This is known as *translational research*.

Translational research is often found in clinical settings, e.g. in cancer research or drug development. But it is also common in cognitive neuroscience.[34] A case in point is the recognition memory study in rats I discussed in chapter 2.3.1.5. Researchers aimed to identify the role of hippocampus in two memory processes both known to be involved in recognition memory: familiarity and recollection. Given what they already knew from research in humans, they aimed to design an experiment in which they would lesion hippocampal structures to see how this affects recognition memory. Since such controlled lesions are not possible in humans, they had to resort to model organisms instead. But to do that, they had to make sure they can use an assessment of recognition performance that mirrors the assessment in humans.

In humans, recognition memory can be assessed by a rather simple paradigm where subjects study lists of words and are subsequently asked if certain words were on the lists they studied. Importantly, though, researchers did not only need subjects' performance but also ratings of how confident they were in remembering or not remembering any given word in the test phase. Only by using performance data together with these confidence ratings could researchers disambiguate whether participants engaged in recollection or familiarity-based recognition processes. Now the tricky question was how to do that in rats. Assessment of performance

---

34 Translation as I describe it here involves transfer between a model and a target system. However, "translational research" is also used to describe cases where transfer occurs between controlled laboratory conditions and clinical practice or from a small model population to larger natural populations—e.g. when developing new kinds of seeds in agriculture oder studying a single patient and translating insights about the functional organization of the brain to humans more generally. Conceived this way, all inference from single cases to populations—and thus much of cognitive neuroscience—would count as translational. The point I want to stress here, however, is not inference from a small number of samples to a general explanations. Rather, I want to emphasize that we can sometimes experiment in one thing that is completely different from, though relevantly similar to, the thing we want to learn about.

is rather straightforward. This just needs researchers to assess whether the rat recognizes an item as learned, and if the rat's recognition or not-recognition of any given item was correct. But how can a rat tell you if it recognizes something? Fortin et al. (2004) trained rats to retrieve a reward in one of two ways depending on whether they recognized an item or not. Rather than words, rats studied pairings of odor and texture presented to them in wells. Basically, the wells were filled with different materials (wool, sand, gravel, etc.) prepared to smell differently (like coffee, vanilla, apple, etc.). A sand-vanilla well, for instance, would be filled with sand and smell like vanilla. Now if the rat encounters a well with a known odor-texture pairing, it will find its reward at the bottom of the well. If it is a new pairing, it will receive a reward at the opposite end of the cage. In order to get the reward, the animal must sniff at the well and decide to either dig or not dig and run to the opposite end of the cage. If it takes the wrong decision, no reward will be received.[35] Once this behavior is learned, and assuming rats are rational reward maximizers, we can use the rats' behavior upon stimulus presentation to indicate whether or not the rat recognized the presented odor-texture pairing. Since we know which pairings an animal was trained on, we also know whether the rat got it right or wrong. This gives us recognition memory performance in rats. But how about the confidence ratings?

Obviously, we cannot ask a rat to rate its confidence by placing a mark on a scale. But we can by varying the risk-benefit ratio. Suppose you are placed in setting where you get a lot in case you are right, nothing in case you are wrong, and a little bit if you just admit that you do not know. Now do you take your chances and guess? This will likely depend on how confident you are about your judgment. If you are very confident, you might as well take the risk and go for the high reward. If you are not confident at all, you may be better off to just pass and take the small reward. If, however, there is only a small reward for being right in the first place and you are not very confident, you might as well take the risk; for you do not have much to loose in the first place. In rats, we can use the same basic rationale: make it harder for the rat to dig by using a higher well so it will only dig into it if it is really worth the effort. In addition, place a default option with a small reward in another part of the cage. Now in case the rat decides to take the default reward (i.e. the rat is unsure if it recognizes the presented odor-texture pairing and does not take its chances), it can neither run to the opposite end of the cage nor dig into the well anymore. If the rat takes the default reward, it has not yet decided whether

---

35 The cage was setup in such a way that once the animal starts digging it cannot get to the other end of the cage anymore. Likewise, once it runs to the opposite end of the cage, it cannot get back to the well-area until the next trial.

or not it recognizes the presented odor-texture pairing. So subsequently, present a well that is a little lower and have the rat go again until it eventually makes a decision to dig or run. This effectively gives us confidence ratings in addition to performance data:[36] the higher the well in the trial where the rat eventually digs or runs, the higher its confidence that it judged the odor-texture pairing correctly as known or unknown (for details see Sauvage 2010).

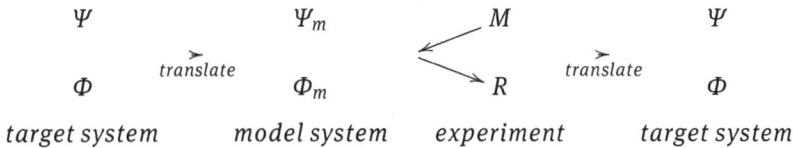

$$\Psi \qquad \Psi_m \qquad\qquad M \qquad\qquad \Psi$$
$$\qquad \xrightarrow{translate} \qquad \Big\backslash \qquad \xrightarrow{translate} \qquad$$
$$\Phi \qquad\quad \Phi_m \qquad\quad R \qquad\qquad \Phi$$

$$\textit{target system} \qquad \textit{model system} \quad \textit{experiment} \qquad \textit{target system}$$

Fig. 14.12: A translational study. The relevant features of the target system/phenomenon (Ψ and Φ) are transferred into a model (Ψ_m, Φ_m) where an experiment is carried out (M represents a manipulation, R the result). Finally, the experimental outcome can be transferred back to the original system/phenomenon.

Now what does this mean with respect to my catalog of experiments? Thus far, my classification of experiments has been neutral with respect to the question where manipulations are carried out—if in a model, a model organism, or the actual target system. This is a virtue of my approach; it can be applied in all of these settings quite independently of whether or not the traget system is being manipulated or observed. But just being able to manipulate different systems and their models is often not enough. In order to exploit experimental manipulations in (computational) models and model organisms we must understand how translational research works and how we can put it to use. We must, that is, sometimes invest a lot of work to develop (computational) models or identify model organisms, localize analogous anatomical structures, adopt paradigms, ensure the reliability of both the experimental method in the model or model organism and the validity of the translational procedure itself, and so on. This is an important and ineliminable

---

36 Obviously, this interpretation of the rat's behavior requires somewhat substantial assumptions; e.g. that rats are natural reward optimizers etc. Considering the validity of such assumptions is important to ensure the applicability of a translational approach. However, to merely illustrate the principle strategy behind translational approaches, we can gloss over the precise assumptions at play in the example and whether or not they are justified.

part of the scientific enterprise. By including translational approaches into my catalog of experiments, I acknowledge this.

While the manipulations we will eventually want to carry out in the (computational) model or model organism might very well be interventionist (or pseudo-interventionist) in character, the translation as such is not. Scientists merely interact with the target system as well as the model or model organism to achieve the best possible translation and prepare the model for intervention: they find the model, develop the task, locate hippocampus in the animal brain, train the animal, etc.[37] Only once this is done will it be useful to actually intervene (in this case selectively lesion hippocampal structures) and observe the effects. And then again, the results obtained from this intervention must be translated back to the target system. A task that, again, does not include interventions but reasoning along the same lines as did translating the paradigm in the first place.

14.11 Upshot

Philosophers of science have recently focused on Woodwardian interventions when discussing scientific experimentation and the construction of mechanistic and/or causal explanations. I have argued that this focus is too short-sighted. There are many ways in which scientists manipulate that do not qualify as Woodward-style interventions. I have called such ways of non-interventionist manipulation *mere interaction*. While mere interactions are united by their not wiggling one thing to see what else wiggles along, there is a wide wide variety of them. In some cases, mere interactions are given an interventionist interpretation. Such *pseudo-interventions*, or as-if-interventions, are an important tool in coming up with scientific explanations. The examples I discussed above illustrate how mere interactions and Woodwardian interventions get employed in many different kinds of experiments; separately or combined. Figure 14.13 summarizes the results of my discussion here.

Note that none of the types of experiments I present here exemplifies a pure intervention study in Woodward's sense (though such cases certainly exist). Yet, all of these experiments make vital contributions to inquiry into causal and constitutive relations. This demonstrates that it is important for contemporary philosophers of science to not gloss over the details of experimental designs by fitting everything

---

[37] Of course, the procedures are usually not as neat and clean as I describe them here. In fact, the translational approach often incurs a lot of additional challenges and complications (see also chapter 9). But for current purposes the scenario I sketched here shall suffice.

into an interventionist picture. Understanding what other kinds of manipulations there are and what these can teach us is essential to understand how practicing scientists come to explain phenomena—in cognitive neuroscience and beyond. Besides, at least in some cases, interventions will not be informative, if at all possible, as long as we do not combine them with mere interactions.

I conclude, therefore, that both Woodward-style interventions and mere interactions—interpreted as pseudo-interventions or not—have their place in the scientific and explanatory enterprise. Both when it comes to causal and other kinds of explanations. The examples I discussed in this chapter nicely illustrate the many different roles that mere interactions can play. They can serve as controlling instances either monitoring or fixing background conditions. They can provide evidence about architectural features of a system. They can illuminate how systems or phenomena develop over time. They can help scientists decompose a phenomenon or system into parts and identify functional units. They can provide information about certain (interlevel) dependencies such as part-whole relations. They can help scientists identify and build models that they can subsequently intervene into. None of this, I take it, would be possible with pure Woodwardian interventions alone.[38] There is much more to empirical manipulations than wiggling some X to see if some Y will wiggle along—and this is not only embellishment, it is essential to how scientists construct explanations! Once we recognize the substantial role that all the different mere interactions play, we find that we suddenly have a handle on those problems that were so troublesome for contemporary "mechanisms and interventions"-style philosophy of science. Most importantly, mere interactions give us a way to identify components before we start to intervene and they help us gain additional evidence about dependence relations such that we can disambiguate between causal and constitutive hypotheses.

Thus far, most of the examples I worked with were rather "simple" experiments from cognitive neuroscience. Once we start considering more complex experimental strategies and designs, I take it, the role of mere interactions will become even more pronounced. For they are important tools that can be used to ensure that interventions worked, hold parameters fixed, or visualize features we could not otherwise see. Also, this catalog will probably grow considerably once we take individual scientific methodologies and their limitations into account. But that will be another project.

---

38 I do not believe Woodward or any proponent of a mechanistic view intend their theories to capture the *only* way in which scientific research proceeds. However, all these different nonwiggling experiments have been largely neglected in contemporary debates. My aim is to bring them into the center of philosophers' attention.

study type	W-I / mi / p-I	what we know	manipulation	what we find	examples
geometric property	mi	system as a whole; what qualifies as entities	staining, cutting, measuring, etc.	size, shape, arrangement, position, location, orientation, connections of entities	histology, gene sequencing
composition	mi	system as a whole, components	putting components together	arrangement of components, relevant components & organization	DNA double helix, C. elegans klinotaxis
time course observation (monitoring)	mi (+ W-I)	system as a whole	apply monitoring apparatus	system behavior (development) unfolding over time	genetic tracers, neural oscillations, plasticity, effects on diff. timescales
model estimation	mi	potentially relevant factors	measure; calculate correlations	predictive model	questionnaires, behavioral studies, etc.
functional units	mi	parts building functional units	measure; calculate correlation matrix	(network of) functional units	resting state functional connectivity
difference association	p-I	groups or conditions	comparison across groups or conditions	associated changes in phenomenon, dissociations	lesions, dissociations, functional imaging
multiple (pseudo-) interventions	many W-Is or p-Is	big part	trigger different functions (repeatedly, alternating), compare across experiments	functional subdivisions	fMRI adaptation
statistical interaction	effectively p-I (main effect as mi, interaction as W-I)	two or more factors relevant for phenomenon	measure, compare across groups and conditions	interaction of factors	2 × 2 design
manipulation check	manipulation is W-I; manipulation check is mi	attempted intervention	assess effect of intervention	if W-I worked	induce affect, administer control questionnaire
translational research	exp. often W-I; translation is mi	phenomenon in target organism	transfer paradigm to model organism	expl. for phenomenon in model organism, transfer to target organism	monkey studies, ROC curves in rats

Fig. 14.13: Table showing the features of different types of experiments. W-I = Woodward-style intervention; mi = mere interaction; p-I = pseudo-intervention, i.e. practically mi but conceptualized as W-I.

15 Conclusions

15.1 Take Home Messages

The major lessons to learn from this book are that there is much more to experimental manipulation than intervention, more to difference making than causation, and more to interlevel manipulation than neat bottom-up and top-down studies. To understand the epistemology behind scientific experiments we must consider a wide range of experimental designs, how scientists use different tools and methodologies, how they analyze and interpret the data they collect, and how they eventually combine this evidence into an explanation. This branch of philosophy of science I call *philosophy of experimentation*.

I set out by asking how scientists explain cognitive phenomena. The answer contemporary philosophy of science offers goes like this: We explain a cognitive phenomenon by identifying its underlying neural *mechanism*. That is, we identify "a set of entities and activities organized such that they exhibit the phenomenon to be explained." (Craver 2007b, p. 5) We do this by systematically *intervening* into the phenomenon, or some component of its mechanism, respectively, and observing the effects. Put briefly, the idea behind these interventions is that we *wiggle* something to see what else will *wiggle along*. If stuff on the *same* (say, neural or cognitive) *level* (or domain) wiggles along, we can infer—given appropriate background conditions, controls, etc.—a *causal* relation. What we intervene into (wiggle) is considered cause of what wiggles along. But we can also manipulate at one level and assess the effect of our manipulation at a different level. If we do this, we will sometimes find interlevel *mutual manipulability* relations. For instance, if we have a mouse perform a spatial memory task and observe activity in hippocampus we intervene at the top (we make the mouse engage in the cognitive process such as spatial orientation) and observe changes at the bottom (we record neural signals). Likewise, we can lesion the mouse's hippocampus (intervene at the bottom) and observe the cognitive impairments (at the top) that result from this lesion. If we find such mutual manipulability relations, proponents of mechanistic explanations argue, this is sufficient to infer a *constitutive* relation. That is, what we wiggle and what wiggles along are related as part and whole rather than as cause and effect.

With this picture, contemporary "mechanisms and interventions"-style philosophy of science laudably reflects the crucial role that manipulability and interlevel experiments play in empirical research. Yet, as I have demonstrated, it is unsatisfying for several reasons. Perhaps most pressingly, mechanisms and interventions cannot be married as easily as is widely believed. This is mainly because there

DOI 10.1515/9783110530940-015

are different kinds of dependence relations in mechanistic explanations: causal and constitutive ones. Problematically, we cannot tease these relations apart by interventionist manipulability. This is partly because interlevel experiments are usually individual unidirectional studies that have a clear temporal dimension and therefore simply *look causal*. Besides, interventions are designed to uncover *causal* relations only and no other dependence relations (such as realization, constitution, or supervenience) may be present among the variables being analyzed. But mechanisms are composed of parts; they non-causally depend on them. Still, mechanists suggest an intervention-based mutual manipulability criterion to identify precisely those constitutive relations between a mechanism as a whole and its component parts. And against the background of empirical practice, it seems that they are well-justified to do so.

To address this problem, I suggested a somewhat weaker reading of Woodward's interventionism: *difference making interventionism* (DMI). The basic idea is that we can modify interventionism's core definitions in such a way that manipulability of some factor Y through intervention into some other factor X does not necessarily indicate a causal relation. Instead, any kind of dependence relation can be indicated. This is to say that if Y wiggles along when we wiggle X, we can conclude that Y *depends* on X, or that X is *relevant* to Y. But based on the observed manipulability relation alone, we do not know whether this is due to a causal, constitutive, or some other kind of dependence between X and Y. Once we adopt this reading, we can happily marry interventionism with mechanistic explanations as we can employ such difference making interventions to uncover general dependencies which include intralevel causal as well as interlevel constitutive ones. Notably, this does not only render interlevel manipulations compatible with mechanistic explanations but also with the common assumptions that cognitive phenomena are *multiply realized* and *supervene* on the neural processes implementing them.

However, even if we adopt DMI, a second problem remains for contemporary "mechanisms and interventions"-style philosophy of science. Before we can even start to intervene and test for causal and/or constitutive relevance relations, we must first delineate phenomena, identify their mechanisms and mechanisms' potential components. This cannot usually be done from the philosophical armchair; it is done by experiment. But experiments serving this purpose do not necessarily employ interventions. Indeed, interventions are something we will usually see somewhere in the middle of the discovery process, not at the very beginning. Or we might not see them at all. Sometimes this is because proper interventions would require us to do things we cannot actually do. In such cases, scientists have their ways of replacing interventions with a whole range of other manipulative, translational, or comparative techniques.

In this context, I introduced the concept of *mere interactions*. Mere interactions are experimental manipulations that do *not* qualify as interventions. Put briefly, they are not *wiggling* one thing to see what else *wiggles along*. In the simplest case, they are best conceived of as *putting something under a magnifying glass*. Applying a stain to a sample of tissue or measuring neural signals using an fMRI machine are cases in point. Mere interactions may involve sophisticated tools (such as an fMRI machine), and they may induce changes (as in the case of staining). But the important point is that they do not manipulate anything for the sake of seeing what the effect of this manipulation will be. Rather, scientists employ mere interactions *because* these manipulations have certain well-known effects. We stain the tissue sample because it will reveal the architecture and arrangement of the cells. But this does not change (given certain idealizing assumptions) the architecture and arrangement of the cells. That has been there *all along*. Similarly, we put participants' heads into an fMRI scanner because we want to see which areas are active. We do not do this to explore how magnetic fields in the fMRI machine affect the brain (at least not, once we have learned that). We do it because we know that the BOLD signals we measure once the brain is inside the magnetic field are indicative of neural activations.

These simple examples already demonstrate that mere interactions play a central role in scientific practice. Like semantics needs to be supplemented with pragmatics to truly understand the meaning of linguistic expressions, and Bayesianism needs Bayesian statistics to get off the ground, we need to supplement interventionism with mere interactions if we want to understand how scientists explain phenomena. Mere interactions allow scientists to observe structures in detail that would not be otherwise accessible, to examine architectural features of systems without destroying them, to study the dynamics of developing systems or complex processes, or to learn about characteristic features of a phenomenon without interference, and so on. Given the role that mere interactions play in empirical science, we may wonder *why* philosophers of science have focused so extensively on Woodwardian interventions when discussing scientific explanations. Partly, I take it, this might be because their traditional project was not to understand experimental practice but to *analyze causal explanatory relations*. However, analyzing what causal explanatory relations are is quite different from understanding how we *discover* them. Besides, we have learned that causation is by far not the only explanatory relation there is. Other dependence relations—such as constitution for instance—can also be explanatory.

Now that we recognize the importance of mere interactions for the scientific explanatory enterprise we can fill the gaps that interventionism and mechanistic explanations left open. Mere interactions give us a way of delineating phenomena and (at least in some cases) a straightforward way of assessing constitutive relations. We can use mere interactions to decompose a system or *zoom into* a mechanism and reveal potential components as well as organizational aspects which we can subsequently subject to further—possibly intervention-based—analyses. Employing mere interactions thus generates evidence that can be used to supplement intervention studies. Without knowing what the implementing mechanism of a phenomenon and its potential components could be, interventions are blind. Besides, observed manipulability alone typically underdetermines the dependence relation at play. Evidence from mere interaction studies can be used to delineate and identify phenomena, their implementing mechanisms and the mechanisms' potential components as well as for disambiguation of different dependency relations (e.g. by assessing part-whole relations through microscopy).

In this context it is worth emphasizing that there are not only clean intervention studies and pure mere interaction studies. In fact, more complex experimental designs will typically employ multiple different manipulations in a single setting. Further, acquired data might be subjected to different analyses to test for different (causal) hypotheses. For instance, certain data sets can even be used to *simulate* interventions. To capture this practice, I introduced the concept of *pseudo-interventions*. Pseudo-interventions are *a way of interpreting sets of mere interactions*. As such, we may consider them as a separate kind of manipulation, although they are reducible to mere interactions. If we compare groups of subjects, say, in a cross-linguistic study, we do not intervene into a person's native language to assess the effects of native language on, say, cortical language processing. In fact, this would be impossible: we cannot simply change a person's native language by intervention. What we can do instead is *monitor* cortical language processing in native speakers of different languages and subsequently compare or *contrast* the data we acquired. This is a case of merely interacting. We can then *interpret* the differences we find as an effect of participants' native language. Though there have not been any interventions at all, our conclusion may read *as if* we performed an intervention experiment when we say, e.g., that different native languages *elicited* differences in cortical language processing.

To fully appreciate the logic behind experimental manipulations we must not only consider studies employing a single manipulation but also take into account more complex experimental designs that employ a combination of different manipulations. For illustration consider a psychologist who is trying to assess the effect of affect on performance in a memory task. In order to assess this, she will induce positive or negative affect, respectively, in her participants and assess their performance before and after this manipulation. While inducing affect may be considered a Woodwardian intervention, the experimental design requires an additional feature: the psychologist has to make sure that her intervention actually induced the desired affect in her participants; for if the intervention did not work in the desired way, the whole experiment will be pointless and she cannot properly evaluate the acquired data. The psychologist can meet this demand by using, say, a questionnaire or doing a simple association test with her participants. This qualifies as a mere interaction: the experimenter interacts with her participants to assess their mood (i.e. whether affect was successfully induced) but she does not manipulate it (at least not during the assessment). Put slightly differently, this is to say that if we want to do Woodwardian wiggling experiments, we must be sure that our wiggling X will actually wiggle X before we can assess its effect on Y. Like delineating phenomena and identifying their underlying mechanisms, this cannot typically be done by interventions. Instead, it requires a manipulation or measurement that allows us to determine whether Y actually wiggles without our assessment causing Y to wiggle. It requires mere interactions, that is. This nicely illustrates the crucial role that mere interactions play in experimental science. Though they may be considered weaker than interventions in some contexts (such as testing causal hypotheses), mere interactions are not in any sense less valuable than interventions. In fact, they (at least sometimes) are the very preconditions or controlling elements that we need for interventions to get off the ground.

For the most part, my discussion here has focused on single experiments or sets of complementary interlevel studies. Note, however, that it is typically not a single experiment that will provide an explanation all by itself. Rather, it typically takes a whole series of experiments using various different methods and manipulations to construct a scientific explanation. The designs these successive studies employ and the evidence they provide must be evaluated in context of other experiments as well as established theories or paradigms. In order to relate different experiments to one another and integrate the evidence they provide into a coherent overall picture, we must think about how various domains, or *levels*, of investigation relate. While I take this to be primarily matter for empirical scientists to figure out rather than a question for metaphysicians, the topic clearly has philosophical significance. For if we talk about interlevel mutual manipulability we must know how to decide whether things are *at the same level* or not. But, as said, I think this is a question we can settle somewhat pragmatically without delving into metaphysics. Given what we see in scientific practice, the following pragmatic criterion of the perspectival view seems feasible: things are considered to be "at the same level" whenever they are *accessible through the same method*. Clearly, there is more to said on this issue, especially on what it means for things to be accessible through the same method. Again though, this is primarily an empirical question, not a metaphysical one. As it stands, this preliminary conception of levels already allows us to examine the basic logic behind interlevel experiments without requiring any major metaphysical commitments.

All in all then, this makes it quite clear that, although interventions and mutual manipulation are powerful tools, they only have a very limited import on actual scientific explanations. Once we consider scientific practice more closely, we will soon realize that the often-cited success stories of how scientists explain cognitive phenomena by (interlevel) interventions draw a skewed, idealized, and heavily polished picture that misses many of the crucial aspects of scientific practice. With the concepts of mere interactions and pseudo-interventions in place, we can re-examine these success stories and re-tell them in a more accurate, less intervention-directed, fashion. Examining the logic behind various different non-interventionist and mixed-manipulation experiments led me to present a whole catalog of experiments that classifies various types of studies with respect to which kinds of manipulations they employ and which research questions they can answer. Thus far, the catalog I provide is not specific to any particular scientific method. But it can be used as a starting point to develop a finer-grained, more detailed catalog that differentiates studies with respect to which precise methodologies they employ and what their individual limitations are, respectively. Yet, already at this early stage the catalog I provide allows us to reassess some of the success stories I outlined in the beginning of this book. As soon as we consider the experimental details of these experiments more carefully, we realize that despite our enthusiasm about bottom-up and top-down interventions there is much much more going on in scientific practice. For instance, Felleman and van Essen's (1991) work on visual processing was based on geometric property studies: they studied histology and connections between different areas rather than carrying out interventions in a Woodwardian sense. Similarly, Ungerleider and Mishkin's (1982) work on visual processing streams is only partly interventional. Lesioning monkeys' brains to see what the effects of these different lesions are qualifies as intervention experiment in a Woodwardian sense. However, comparing across groups to dissociate visual processing streams is only pseudo-interventional. The same holds true for the many different imaging studies discussed in the context of memory and language research, respectively. For the most part, these experiments were actually pseudo-interventional difference association studies, rather than genuine intervention studies. Similarly, the EEG studies discussed in the context of memory consolidation (see section 2.3.1.4) mostly exemplify time course observation studies. For instance, the hypothesis that information learned during wakefulness is continuously replayed during sleep (cf. Buzsaki & Draguhn 2004) is mostly fueled by observations of oscillatory patterns over time. Likewise, functionally specific cortical oscillations have been identified mostly by observing these oscillations while participants engaged in specific tasks (e.g. Axmacher et al. 2006; Thut & Miniussi 2009); these, again, are instances of time course observation studies rather than intervention studies.

15.2 Wrapping Up

Closely considering the methodology behind different empirical strategies, I have illuminated the first steps on a path towards an epistemology of scientific experimentation that goes beyond contemporary "mechanisms and interventions"-style philosophy of science. By taking this path, I suggest, we can avoid some of the major challenges contemporary philosophy of science faces and learn that there is much more to scientific investigation than the popular wiggling conception of scientific experimentation suggests. The answer to the question of how scientists explain (cognitive) phenomena lies in the precise materials and methods of empirical research.

The critic may object that I have merely underlined the fact that scientific observation is theory-driven. My reply to this worry is that I have said much more than this. I have demonstrated that scientific observation is not just driven by theories but also by available tools and methodologies. I have illustrated that different kinds of experiments and different ways of data analysis provide answers to different research questions. Besides, experimental manipulations must always be viewed in context of other manipulations, background conditions, etc. rather than just in a theoretical context. In the end, it is a somewhat sophisticated systematic combination of various different manipulative and observational techniques that leads scientists to discover bits of explanatory information and integrate them into a coherent picture.

To understand how scientific explanations are constructed, therefore, we must understand the dynamics of experimental research and the logic behind scientific experiments. The catalog I provide here is a start.

Admittedly, none of this is going to cure cancer. But it will help us to better understand how scientists explain (not only) cognitive phenomena, and how they find the causes of diseases and discover possible cures. And maybe, just maybe, a systematic examination of different experimental strategies and designs will inspire future experiments and thereby contribute to scientific explanations and discoveries.

References

Andersen, H. (2014). A field guide to mechanisms. *Philosophy Compass*, 9, 274–293.

Axmacher, N., Mormann, F., Fernández, G., Elger, C. E., & Fell, J. (2006). Memory formation by neuronal synchronization. *Brain Research Reviews*, 52, 170–182.

Bacon, F. (1960 [1620]). *The New Organon*. New York, NY: Bobbs-Merrill.

Baddeley, A., Eysenck, M. W., & Anderson, M. (2009). *Memory*. Hove: Psychology Press.

Baumgartner, M. (2010). Interventionism and epiphenomenalism. *Canadian Journal of Philosophy*, 40, 359–384.

Baumgartner, M. (2013). Rendering interventionism and non-reductive physicalism compatible. *Dialectica*, 67, 1–27.

Baumgartner, M. & Casini, L. (forthcoming). An abductive theory of constitution. *Philosophy of Science*.

Bechtel, W. (2008). *Mental Mechanisms*. London and New York: Routledge.

Bechtel, W. (2011). Mechanism and biological explanation. *Philosophy of Science*, 78, 533–557.

Bechtel, W. & Abrahamsen, A. (2005). Explanation: a mechanist alternative. *Studies in History and Philosophy of Biological and Biomedical Sciences*, 36, 421–441.

Bechtel, W. & Abrahamsen, A. (2008). From reduction back to higher levels. In: B. C. Love, K. McRae, & V. M. Sloutsky (eds.), *Proceedings of the 30th Annual Conference of the Cognitive Science Society*, Austin: Cognitive Science Society, pp. 559–564.

Bechtel, W., Abrahamsen, A., & Graham, G. (1999). The life of cognitive science. In: W. Bechtel & G. Graham (eds.), *A Companion to Cognitive Science*, Malden, MA: Blackwell, pp. 1–104.

Bechtel, W. & Richardson, R. (1993). *Discovering complexity. Decomposition and localization as strategies in scientific research*. Princeton, NJ: Princeton University Press.

Beckermann, A. (2001). *Analytische Einführung in die Philosophie des Geistes*. Berlin: DeGruyter.

Bennett, K. (2011). Construction area (no hard hat required). *Philosophical Studies*, 154, 79–104.

Berger, H. (1929). Über das Elektrenkephalogramm des Menschen. *Archiv für Psychiatrie und Nervenkrankheiten*, 87, 527–550.

Bermúdez, J. (2003). *Thinking Without Words*. Oxford: Oxford University Press.

Bermúdez, J. L. (2010). *Cognitive Science*. Cambridge, MA: Cambridge University Press.

Bickle, J. (1998). *Psychoneural Reduction: The New Wave*. Cambridge, MA: MIT Press.

Bickle, J. (2003). *Philosophy and neuroscience: A ruthlessly reductive account*. Dordrecht: Kluwer Academic.

Bickle, J. (2006). Reducing mind to molecular pathways: explicating the reductionism implicit in current cellular and molecular neuroscience. *Synthese*, 151, 411–434.

Blumstein, S., Cooperc, W., Zurife, E., & Caramazza, A. (1977). The perception and production of voice onset time in aphasia. *Neuropsychologia*, 15, 371–383.

Boden, M. (2007). The history of cognitive science: seven key dates. In: S. Vosniadou, D. Kayser, & A. Protopapas (eds.), *Proceedings of the European Cognitive Science Conference 2007*. London: Psychology Press, pp. 4–9.

Boone, W. & Piccinini, G. (2016). Mechanistic abstraction. *Philosophy of Science*, 83, 686–697.

Broad, C. D. (1925). *The Mind and its Place in Nature*. London: Kegan Paul, Trench, Trübner & Co.

Broca, P. (1861). Remarques sur le siége de la faculté du langage articulé suives d'une observation d'aphémie. *Bulletin et Memoires de la Societe Anatomique de Paris*, 2, 330–357.

Buzsaki, G. & Draguhn, A. (2004). Neuronal oscillations in cortical networks. *Science*, 304, 1926–1929.

DOI 10.1515/9783110530940-016

Campagner, R. & Galavotti, M. (2007). Plurality in causality. In: W. G. & M. P. (eds.), *Thinking about causes: from Greek philosophy to modern physics.*, Pittsburgh, PA: University of Pittsburgh Press, pp. 178–199.

Campbell, R., MacSweeney, M., & Waters, D. (2007). Sign language and the brain: a review. *Journal of Deaf Studies and Deaf Education*, 13, 3–20.

Capek, C. M., Waters, D., Woll, B., MacSweeney, M., Brammer, M. J., McGuire, P. K., David, A. S., & Campbell, R. (2008). Hand and mouth: Cortical correlates of lexical processing in british sign language and speechreading english. *Journal of Cognitive Neuroscience*, 20, 1220–1234.

Cardin, V., Orfanidou, E., Kästner, L., Rönnberg, J., Woll, B., Capek, C., & Rudner, M. (2016). Monitoring different phonological parameters of sign language engages the same cortical language network but distinctive perceptual ones. *Journal of Cognitive Neuroscience*, 28, 20–40.

Cartwright, N. (2002). Against modularity, the causal markov condition and any link between the two. *British Journal for the Philosophy of Science*, 53, 411–453.

Casini, L., McKay Illari, P., Russo, F., & Williamson, J. (2011). Models for prediction, explanation and control: Recursive bayesian networks. *Theoria*, 70, 5–33.

Clark, A. (1998). Magic words: How language augments human computation. In: P. Carruthers & J. Boucher (eds.), *Language and Thought: Interdisciplinary Themes*, Cambridge: Cambridge University Press, pp. 162–183.

Clark, S. A., Allard, T., Jenkins, W. M., & Merzenich, M. M. (1988). Receptive fields in the body-surface map in adult cortex defined by temporally correlated inputs. *Nature*, 332, 444–445.

Clarke, B., Leuridan, B., & Williamson, J. (2014). Modeling mechanisms with causal cycles. *Synthese*, 191, 1651–1681.

Corina, D. P., Jose-Robertson, L. S., Guillemin, A., High, J., & Braun, A. R. (2003). Language lateralization in a bimanual language. *Journal of Cognitive Neuroscience*, 15, 718–730.

Corina, D. P., McBurney, S. L., Dodrill, C., Hinshaw, K., Brinkley, J., & Ojeman, G. (1999). Functional roles of broca's area and smg: evidence from cortical stimulation mapping in a deaf signer. *Neuroimage*, 10, 570–581.

Couch, M. (2011). Mechanisms and constitutive relevance. *Synthese*, 183, 375–388.

Craver, C. F. (2002). Interlevel experiments and multilevel mechanisms in the neuroscience of memory. *Philosophy of Science (Suppl.)*, 69, S83–S97.

Craver, C. F. (2006). When mechanistic models explain. *Synthese*, 153, 355–376.

Craver, C. F. (2007a). Constitutive explanatory relevance. *Journal of Philosophical Research*, 32, 3–20.

Craver, C. F. (2007b). *Explaining the Brain: Mechanisms and the Mosaic Unity of Neuroscience.* New York: Oxford University Press.

Craver, C. F. (2013). Functions and mechanisms: A perspectivalist view. *Synthese*, 363, 133–158.

Craver, C. F. (2015). Levels. *Open MIND*, 8, 1–26.

Craver, C. F. & Bechtel, W. (2007). Top-down causation without top-down causes. *Biology and Philosophy*, 22, 547–563.

Craver, C. F. & Darden, L. (2001). Discovering mechanisms in neurobiology: The case of spatial memory. In: P. Machamer, R. Grush, & P. McLaughlin (eds.), *Theory and Method in Neuroscience*, Pittsburgh, PA: University of Pittsburgh Press, pp. 112–137.

Craver, C. F. & Darden, L. (2005). Introduction. *Studies in History and Philosophy of Biological and Biomedical Sciences*, 36, 233–244.

Craver, C. F. & Darden, L. (2013). *In Search of Mechanisms: Discoveries Across the Life Sciences*. Chicago, IL: University of Chicago Press.

Craver, C. F. & Kaplan, D. (2016). Are more details better? On the norms of completeness for mechanistic explanations. Manuscript.

Craver, C. F. & Tabery, J. (2015). Mechanisms in science. In: E. Zalta (ed.), *The Stanford Encyclopedia of Philosophy (Spring 2016 Edition)*, URL = <http://plato.stanford.edu/archives/spr2016/entries/science-mechanisms/>.

Darden, L. (2006). *Reasoning in Biological Discoveries*. Oxford: Oxford University Press.

Davis, M. & Gaskell, M. (2009). A complementary systems account of word learning: neural and behavioral evidence. *Philosophical Transactions of the Royal Society B*, 364, 3773–3800.

Davis, M. & Johnsrude, I. (2003). Hierarchical processing in spoken language comprehension. *Journal of Neuroscience*, 23, 3423–3431.

de Bruin, L. & Kästner, L. (2012). Dynamic embodied cognition. *Phenomenology and the Cognitive Sciences*, 11, 541–563.

Dennett, D. (1991). Real patterns. *Journal of Philosophy*, 88, 27–51.

Dick, F., Lee, H. L., Nusbaum, H., & Price, C. J. (2011). Auditory-motor expertise alters 'speech selectivity' in professional musicians and actors. *Cerebral Cortex*, 21, 938–948.

Dretske, F. (1993). Mental events as structuring causes of behavior. In: J. Heil & A. Mele (eds.), *Mental causation*, Oxford: Clarendon Press.

Dupré, J. (2013). Living causes. *Proceedings of the Aristotelian Society*, Suppl. Vol. 87, 19–38.

Düzel, E., Penny, W., & Burgess, N. (2010). Brain oscillations and memory. *Current Opinion in Neurobiology*, 20, 143–149.

Eichenbaum, H. & Cohen, N. J. (2004). *From Conditioning to Conscious Recollection: Memory Systems of the Brain*. New York, NY: Oxford University Press.

Emmorey, K. (2002). *Language, Cognition, and the Brain. Insights from Sign Language Research*. Mahwah, NJ: Lawrence Erlbaum.

Emmorey, K., Grabowski, T., McCullough, S., Damasio, H., Ponto, L., Hichwa, R., & Bellugi, U. (2003). Neural systems underlying lexical retrieval for sign language. *Neuropsychologia*, 41, 85–95.

Emmorey, K., Mehta, S., & Grabowski, T. J. (2007). The neural correlates of sign versus word production. *Neuroimage*, 36, 202–208.

Eronen, M. I. (2010). *Reduction in Philosophy of Mind: a Pluralistic Account*. Ph.D. thesis, University of Osnabrück.

Eronen, M. I. (2012). Pluralistic physicalism and the causal exclusion argument. *European Journal for Philosophy of Science*, 2, 219–232.

Eronen, M. I. (2013). No levels, no problems: Downward causation in neuroscience. *Philosophy of Science*, 80, 1042–1052.

Eronen, M. I. (2015). Levels of organization: A deflationary account. *Biology and Philosophy*, 30, 39–58.

Fazekas, P. & Kertész, G. (2011). Causation at different levels: tracking the commitments for mechanistic explanations. *Biology and Philosophy*, 26, 365–383.

Feest, U. (2016). Phenomena and objects of research in the cognitive and behavioral sciences. Paper presented at 25th Biennial Meeting of the Philosophy of Science Association, Nov 3–5 2016, Atlanta, GA, USA.

Felleman, D. J. & Essen, D. C. V. (1991). Distributed hierarchical processing in the primate cerebral cortex. *Cerebral Cortex*, 1, 1–47.

Field, A. (2009). *Discovering Statistics Using SPSS*. London: SAGE Publications, 3rd ed.

Fodor, J. A. (1975). *The Language of Thought*. Cambridge, MA: Harvard University Press.

Fortin, N. J., Wright, S. P., & Eichenbaum, H. (2004). Recollection-like memory retrieval in rats is dependent on the hippocampus. *Nature*, 431, 188–191.

Franklin-Hall, L. (2016a). High-level explanation and the interventionist's 'variables problem'. *British Journal for the Philosophy of Science*, 67, 553–577.

Franklin-Hall, L. (2016b). New mechanistic explanation and the need for explanatory constraints. In: K. Aizawa & C. Gillett (eds.), *Scientific Composition and Metaphysical Ground*, London: Palgrave Macmillan, chap. 2, pp. 41–74.

Friederici, A. (2002). Towards a neural basis of auditory sentence processing. *Trends in Cognitive Sciences*, 6, 78–84.

Frishberg, N. (1975). Arbitrariness and iconicity: historical change in american sign language. *Language*, 51, 940–961.

Gazzaniga, M., Ivry, R., & Mangun, G. (2008). *Cognitive Neuroscience: The Biology of the Mind*. New York, NY: W.W. Norton and Company., 3rd ed.

Gebharter, A. (2014). A formal framework for representing mechanisms? *Philosophy of Science*, 81, 138–153.

Gebharter, A. (2015). Causal exclusion and causal Bayes nets. *Philosophy and Phenomenological Research*, doi: 10.1111/phpr.12247.

Gebharter, A. (2016). Uncovering constitutive relevance relations in mechanisms. *Philosophical Studies*, doi: 10.1007/s11098-016-0803-3.

Gebharter, A. & Baumgartner, M. (2016). Constitutive relevance, mutual manipulability, and fat-handedness. *British Journal for the Philosophy of Science*, 67, 731–756.

Gebharter, A. & Kaiser, M. I. (2014). Causal graphs and biological mechanisms. In: M. I. Kaiser, O. Scholz, D. Plenge, & A. Hüttemann (eds.), *Explanation in the special sciences: The case of biology and history*, Dordrecht: Springer, pp. 55–86.

Geschwind, N. (1967). The varieties of naming errors. *Cortex*, 3, 97–112.

Giere, R. (2006). *Scientific Perspectivism*. Chicago, IL: University of Chicago Press.

Gijsbers, V. & de Bruin, L. (2014). How agency can solve interventionism's problem of circularity. *Synthese*, 191, 1775–1791.

Gillett, C. (2010). Moving beyond the subset model of realization: The problem of qualitative distinctness in the metaphysics of science. *Synthese*, 177, 165–192.

Gillett, C. (2011). Multiply realizing scientific properties and their instances. *Philosophical Psychology*, 24, 727–738.

Glennan, S. (1996). Mechanisms and the nature of causation. *Erkenntnis*, 44, 49–71.

Glennan, S. (2002). Rethinking mechanistic explanation. *Philosophy of Science*, 69, S342–S353.

Glennan, S. (2005). Modeling mechanisms. *Studies in History and Philosophy of Biological and Biomedical Sciences*, 36, 443–464.

Glennan, S. (2009). Productivity, relevance and natural selection. *Biology and Philosophy*, 24, 325–339.

Glennan, S. (2010a). Ephemeral mechanisms and historical explanation. *Erkenntnis*, 72, 251–266.

Glennan, S. (2010b). Mechanisms, causes, and the layered model of the world. *Philosophy and Phenomenological Research*, 81, 362–381.

Glennan, S. (2013). Mechanistic constitution and difference making. Presentation at Universität zu Köln.

Godfrey-Smith, P. (2012). Causal pluralism. In: H. Beebee, C. Hitchcock, & P. Menzies (eds.), *Oxford handbook of causation*, New York, NY: Oxford University Press, pp. 326–338.

Goldstein, E. (2004). *Cognitive Psychology: Connecting Mind, Research and Everyday Experience.* Belmont, CA: Wadsworth Publishing.

Goodale, M. & Milner, A. (1992). Separate visual pathways for perception and action. *Trends in Neurosciences*, 15, 20–25.

Grinvald, A. & Hildesheim, R. (2004). VSDI: A new era in functional imaging of cortical dynamics. *Nature Reviews Neuroscience*, 5, 874–885.

Gruber, T., Tsivilis, D., Giabbiconi, C.-M., & Müller, M. (2008). Induced electroencephalogram oscillations during source memory: familiarity is reflected in the gamma band, recollection in the theta band. *Journal of Cognitive Neuroscience*, 20, 1043–53.

Guderian, S. & Düzel, E. (2005). Induced theta oscillations mediate large-scale synchrony with mediotemporal areas during recollection in humans. *Hippocampus*, 15, 901–912.

Hall, N. (2004). Two concepts of causation. In: J. Collins, N. Hall, & L. A. Paul (eds.), *Causation and Counterfactuals*, Cambridge, MA: MIT Press.

Harbecke, J. (2010). Mechanistic constitution in neurobiological explanations. *International Studies in the Philosophy of Science*, 24, 267–285.

Harbecke, J. (2013). Regularity theories of mechanistic constitution in comparison. In: W. S. M. Hoeltje, T. Spitzley (ed.), *GAP.8 Was können wir glauben? Was dürfen wir tun?*, Duisburg-Essen: DuEPublico, pp. 126–134.

Harbecke, J. (2015). Regularity constitution and the location of levels. *Foundations of Science*, 20, 3, 323–338.

Harinen, T. (2014). Mutual manipulability and causal inbetweenness. *Synthese*, doi: 10.1007/s11229-014-0564-5.

Hausman, D. M. & Woodward, J. (1999). Independence, invariance and the Causal Markov Condition. *British Journal for the Philosophy of Science*, 50, 521–583.

Haxby, J. V., Horwitz, B., Ungerleider, L. G., Maisog, J. M., Pietrini, P., & Grady, C. L. (1994). The functional organization of human extrastriate cortex: a PET-rCBF study of selective attention to faces and locations. *Journal of Neuroscience*, 14, 6336–6353.

Hebb, D. O. (1949). *The Organization of Behavior.* New York: Wiley.

Hempel, C. & Oppenheim, P. (1948). Studies in the logic of explanation. *Philosophy of Science*, 15, 135–175.

Hickok, G. & Poeppel, D. (2000). Towards a functional neuroanatomy of speech perception. *Trends in Cognitive Sciences*, 4, 131–138.

Hickok, G. & Poeppel, D. (2004). Dorsal and ventral streams: A framework for understanding aspects of the functional anatomy of language. *Cognition*, 91, 67–99.

Hickok, G. & Poeppel, D. (2007). The cortical organization of speech processing. *Nature Reviews Neuroscience*, 8, 393–402.

Hitchcock, C. (2007). How to be a causal pluralist. In: G. Wolters & P. Machamer (eds.), *Thinking about causes: from Greek philosophy to modern physics*, Pittsburgh, PA: University of Pittsburgh Press, pp. 200–221.

Hoffmann, R. & Deffenbacher, K. (1992). A brief history of applied cognitive psychology. *Applied Cognitive Psychology*, 6, 1–48.

Hoffmann-Kolss, V. (2014). Interventionism and higher-level causation. *International Studies in the Philosophy of Science*, 28, 49–64.

Hubel, D. H. & Wiesel, T. N. (1959). Receptive fields of single neurons in the cat's striate cortex. *Journal of Physiology*, 148, 574–591.

Hume, D. (1902 [1777]). *Enquiries concerning Human understanding and concerning the principles of morals.* Oxford: Oxford University Press, 2nd ed.

Illari, P. (2013). Mechanistic explanation: integrating the ontic and the epistemic. *Erkenntnis*, 78, 237–255.

Illari, P., Russo, F., & Williamson, J. (2011). Why look at causality in the sciences? A manifesto. In: P. McKay Illari, F. Russo, & J. Williamson (eds.), *Causality in the Sciences*, Oxford University Press, pp. 3–22.

Illari, P. & Williamson, J. (2012). What is a mechanism? Thinking about mechanisms across the sciences. *European Journal of Philosophy of Science*, 2, 119–135.

Izquierdo, E. J. & Beer, R. D. (2013). Connecting a connectome to behavior: An ensemble of neuroanatomical models of c. elegans klinotaxis. *PLOS Computational Biology*, 9, e1002890.

Jacob, F. & Monod, J. (1961). Genetic regulatory mechanisms in the synthesis of proteins. *Journal of Molecular Biology*, 3, 318–356.

Jacoby, L. & Dallas, M. (1981). On the relationship between autobiographical memory and perceptual learning. *Journal of Experimental Psychology: General*, 3, 306–340.

Jenkins, W., Merzenich, M., Ochs, M., Allard, T., & Guic-Robles, E. (1990). Functional reorganization of primary somatosensory cortex in adult owl monkeys after behaviorally controlled tactile stimulation. *Journal of Neurophysiology*, 63, 82–104.

Kaiser, M. I. & Krickel, B. (2016). The metaphysics of constitutive mechanistic phenomena. *British Journal for the Philosophy of Science*, doi: 10.1093/bjps/axv058.

Kandel, E. R. (2006). *In Search of Memory: The Emergence of a New Science of Mind*. New York, NY: W. W. Norton & Company.

Kaneld, E. R., Schwartz, J. H., & Jessell, T. M. (2000). *Principles of Neural Science*. Blacklick, OH: Mcgraw-Hill Professional, 4 ed.

Kaplan, D. M. (2011). Explanation and description in computational neuroscience. *Synthese*, 183, 339–373.

Kaplan, D. M. & Bechtel, W. (2011). Dynamical models: An alternative or complement to mechanistic explanations? *Topics in Cognitive Science*, 3, 438–444.

Kaplan, D. M. & Craver, C. F. (2011). The explanatory force of dynamical models. *Philosophy of Science*, 78, 601–627.

Kästner, L. (2009). What is cognition? Bachelor Thesis. University of Osnabrück.

Kästner, L. (2010a). How, and where, do you know it? on different accounts of semantic memory. Manuscript.

Kästner, L. (2010b). *"What" & "How"/"Where" in the Signing Brain*. Master's thesis, University College London.

Kästner, L. (2011). Mechanistic explanation and interventionism: An untruthful marriage. Paper presented at the (E)SPP 2011.

Kästner, L. (2013). Mechanistic explanation & interlevel manipulation. Paper presented at the SPP 2013.

Kästner, L. (2015). Learning about constitutive relations. In: U. Mäki, S. Ruphy, G. Schurz, & I. Votsis (eds.), *EPSA13 Helsinki, European Studies in Philosophy of Science*, Springer, vol. 1, pp. 155–167.

Kästner, L. (2016). The mechanistic triad: Produce, underlie and maintain. Manuscript.

Kästner, L. (2017). Making sense of causation in psychiatry. Manuscript.

Kästner, L. (in press). Levels as epistemic perspectives. *Studies in the History and Philosophy of Science*.

Kästner, L. & Andersen, L. M. (in press). Intervening into mechanisms: Prospects and challenges. *Philosophy Compass*.

Kästner, L. & Haueis, P. (2017). Mechanisms, patterns and the ontic/epistemic distinction. Manuscript.

Kästner, L. & Walter, S. (2013). Historical perspectives on the what and the where of cognition. In: C. Pléh, L. Gurova, & L. Ropolyi (eds.), *New Perspectives on the History of Cognitive Science*, Budapest: Akadémiai Kiadò.

Kim, J. (1998). *Mind in a Physical World*. Cambridge, MA: MIT Press.

Kim, J. (2005). *Physicalism or Something Near Enough*. Princeton, NJ: Princeton University Press.

Kirov, R., Weiss, C., Siebner, H., Born, J., & Marshall, L. (2009). Slow oscillation electrical brain stimulation during waking promotes EEG theta activity and memory encoding. *Proceedings of the National Academy of Sciences*, 106, 15460–15465.

Kistler, M. (2007). Mechanisms and downward causation. *EPSA07: 1st Conference of the European Philosophy of Science Association*.

Kitcher, P. (1984). 1953 and all that: a tale of two sciences. *Philosophical Review*, 93, 335–373.

Krekelberg, B., Boynton, G. M., & van Wezel, R. J. (2006). Adaptation: from single cells to BOLD signals. *Trends in Neurosciences*, 9, 250–256.

Krickel, B. (2014). *The Metaphysics of Mechanisms*. Ph.D. thesis, Humboldt-Universität zu Berlin.

Krickel, B. (2017). Constitutive relevance: what it is and how it can be defined in terms of interventionism. Manuscript.

Kumari, V., Peters, E. R., Fannon, D., Antonova, E., Premkumar, P., Anilkumar, A. P., Williams, S. C., & Kuipers, E. (2009). Dorsolateral prefrontal cortex activity predicts responsiveness to cognitive–behavioral therapy in schizophrenia. *Biological Psychiatry*, 66, 594–602.

Kuorikoski, J. & Ylikoski, P. (2013). How organization explains. In: D. Dieks & V. Karakostas (eds.), *EPSA11 Perspectives and Foundational Problems in Philosophy of Science*, Dordrecht: Springer, pp. 69–80.

Lambon Ralph, M. A., Lowe, C., & Rogers, T. T. (2007). Neural basis of category-specific semantic deficits for living things: evidence from semantic dementia, HSVE and a neural network model. *Brain*, 130, 1127–1137.

Leech, R., Holt, L., Devlin, J., & Dick, F. (2009). Expertise with artificial nonspeech sounds recruits speech-sensitive cortical regions. *Journal of Neuroscience*, 29, 5234–5239.

Leuridan, B. (2012). Three problems for the mutual manipulability account of constitutive relevance in mechanisms. *The British Journal for the Philosophy of Science*, 63, 399–427.

Liberman, A. & Wahlen, D. (2000). On the relation of speech to language. *Trends in Cognitive Sciences*, 4, 187–196.

Lindemann, B. (2010). Are top-down experiments possible? Manuscript.

Lingnau, A., Gesierich, B., & Caramazza, A. (2009). Asymmetric fMRI adaptation reveals no evidence for mirror neurons in humans. *Proceedings of the National Academy of Sciences of the United States of America*, 106, 9925–9930.

List, C. & Menzies, P. (2009). Nonreductive physicalism and the limits of the exclusion principle. *Journal of Philosophy*, 106, 475–502.

Lømo, T. (2003). The discovery of long-term potentiation. *Philosophical Transactions of the Royal Society B*, 358, 617–620.

Machamer, P. (2004). Activities and causation: The metaphysics and epistemology of mechanisms. *International Studies in the Philosophy of Science*, 18, 27–39.

Machamer, P. K., Darden, L., & Craver, C. F. (2000). Thinking about mechanisms. *Philosophy of Science*, 67, 1–25.

Mackie, J. L. (1974). *The cement of the universe*. New York, NY: Oxford University Press.

MacSweeney, M., Campbell, R., Woll, B., Giampietro, V., David, A., McGuire, P., Calvert, G., & Bramers, M. (2004). Dissociating linguistic and nonlinguistic gestural communication in the brain. *Neuroimage*, 22, 1605–1618.

MacSweeney, M., Capek, C. M., Campbell, R., & Woll, B. (2008). The signing brain: the neurobiology of sign language. *Trends in Cognitive Sciences*, 12, 432–440.

MacSweeney, M., Woll, B., Campbell, R., Calvert, G. A., McGuire, P. K., David, A. S., Simmons, A., & Brammer, M. J. (2002). Neural correlates of British Sign Language comprehension: spatial processing demands of topographic language. *Journal of Cognitive Neuroscience*, 14, 1064–1075.

Maguire, E. A., Woollett, K., & Spiers, H. J. (2006). London taxi drivers and bus drivers: A structural MRI and neuropsychological analysis. *Hippocampus*, 16, 1091–1101.

Manns, J. R., Hopkins, R. O., Reed, J. M., Kitchener, E. G., & Squire, L. R. (2003). Recognition memory and the human hippocampus. *Neuron*, 37, 171–180.

Marcellesi, A. (2011). Manipulation and interlevel causation. Manuscript.

Marcellesi, A. (2013). Invariance and causal explanation: Missed connections. Manuscript.

Marr, D. (1982). *A Computational Investigation Into the Human Representation and Processing of Visual Information*. New York, NY: W.H. Freeman.

Mc Manus, F. (2012). Development and mechanistic explanation. *Studies in History and Philosophy of Biological and Biomedical Sciences*, 43, 532–541.

McCulloch, W. S. & Pitts, W. (1943). A logical calculus of the ideas immanent in nervous activity. *Bulletin of Mathematical Biology*, 5, 4, 115–133.

McGuire, P. K., Robertson, D., Thacker, A., David, A. S., Kitson, N., Frackowiak, R. S. J., & Frith, C. D. (1997). Neural correlates of thinking in sign language. *Neuroreport*, 8, 695–698.

McLaughlin, B. & Bennett, K. (2011). Supervenience. In: E. Zalta (ed.), *The Stanford Encyclopedia of Philosophy (Spring 2016 Edition)*, URL = <https://plato.stanford.edu/entries/supervenience/>.

Meier, I. & Sandler, W. (2004). *Language in Space: A Window on Israeli Sign Language (Hebrew)*. Haifa: University of Haifa Press.

Menzies, P. (2008). The exclusion problem, the determination relation, and contrastive causation. In: J. Hohwy & J. Kallestrup (eds.), *Being Reduced: New Essays on Reduction, Explanation, and Causation*, Oxford: Oxford University Press, pp. 196–217.

Merzenich, M., Nelson, R., Stryke, M., Cynader, M., Schoppmann, A., & Zook, J. (1984). Somatosensory cortical map changes following digit amputation in adult monkeys. *Journal of Comparative Neurology*, 224, 591–605.

Miller, G. (2003). The cognitive revolution: a historical perspective. *Trends in Cognitive Science*, 7, 141–144.

Milner, B., Corkin, S., & Teuber, H. L. (1968). Further analysis of the hippocampal amnesic syndrome: 14 year follow-up study of H.M. *Neuropsychologia*, 6, 215–234.

Minsky, M. (1988). *The Society of Mind*. New York, NY: Simon & Schuster.

Mitchell, S. (2002). Integrative pluralism. *Biology and Philosophy*, 17, 55–70.

Mitchell, S. (2003). *Biological Complexity and Integrative Pluralism*. Cambridge: Cambridge University Press.

Morgan, M. (2003). Experiments without material intervention. In: H. Radder (ed.), *The Philosophy of Scientific Experimentation*, Pittsburgh, PA: University of Pittsburgh Press, chap. 11, pp. 216–235.

Morris, R., Garrud, P., Rawlings, J., & O'Keefe, J. (1982). Place navigation impaired in rats with hippocampal lesion. *Nature*, 297, 681–683.

Neisser, U. (1967). *Cognitive Psychology*. New York, NY: Appleton-Century-Crofts.

Neville, H. J., Bavelier, D., Corina, D., Rauschecker, J., Karni, A., Lalwani, A., Braun, A., Clark, V., Jezzard, P., & Turner, R. (1998). Cerebral organization for language in deaf and hearing subjects: biological constraints and effects of experience. *Proceedings of the National Academy of Sciences of the United States of America*, 95, 922–929.

Newell, A., Shaw, J. C., & Simon, H. (1958). Elements of a theory of human problem solving. *Psychological Review*, 65, 3, 151–166.

Newen, A. (2013). *Philosophie des Geistes*. München: C. H. Beck.

O'Keefe, J. & Dostrovsky, J. (1971). The hippocampus as a spatial map. Preliminary evidence from unit activity in the freely-moving rat. *Brain Research*, 34, 171–175.

Olton, D. S., Branch, M., & Best, P. J. (1978). Spatial correlates of hippocampal unit activity. *Experimental Neurology*, 58, 387–409.

Patterson, K., Nestor, P., & Rogers, T. (2007). Where do you know what you know? The representation of semantic knowledge in the human brain. *Nature Reviews Neuroscience*, 8, 976–987.

Pauen, M. & Stephan, A. (eds.) (2002). *Phänomenales Bewusstsein - Rückkehr zur Identitätstheorie?* Paderborn: Mentis.

Pavlov, I. (1960 [1927]). *Conditional Reflexes*. New York: Dover Publications.

Pearl, J. (2000). *Causality: models, reasoning, and inference*. Cambridge, MA: Cambridge University Press, 2nd ed.

Petitto, L. A., Zatorre, R. J., Gauna, K., Nikelski, E. J., Dostie, D., & Evans, A. C. (2000). Speech-like cerebral activity in profoundly deaf people processing signed languages: Implications for the neural basis of human language. *Proceedings of the National Academy of Sciences of the United States of America*, 97, 13961–13966.

Piccinini, G. (2004). The first computational theory of mind and brain: a close look at McCulloch and Pitts's 'logical calculus of ideas immanent in nervous activity'. *Synthese*, 141, 175–215.

Piccinini, G. & Bahar, S. (2013). Neural computation and the computational theory of cognition. *Cognitive Science*, 37, 453–488.

Place, U. (1956). Is consciousness a brain process. *British Journal of Psychology*, 47, 44–50.

Plihal, W. & Born, J. (2008). Effects of early and late nocturnal sleep on declarative and procedural memory. *Journal of Cognitive Neuroscience*, 94, 534–547.

Polger, T. (2010). Mechanisms and explanatory realization relations. *Synthese*, 177, 193–212.

Polger, T. & Shapiro, L. (2008). Understanding the dimensions of realization. *Journal of Philosophy*, 105, 213–222.

Potochnik, A. (2010). Levels of explanation reconceived. *Philosophy of Science*, 77, 59–72.

Poulin-Dubois, D., Sodian, B., Metz, U., Tilden, J., & Schoeppner, B. (2007). Out of sight is not out of mind: Developmental changes in infants' understanding of visual perception during the second year. *Journal of Cognition and Development*, 8, 401–421.

Power, J. D., Cohen, A. L., Nelson, S. M., Wig, G. S., Barnes, K. A., Church, J. A., Vogel, A. C., Laumann, T. O., Miezin, F. M., Schlaggar, B. L., & Petersen, S. E. (2011). Functional network organization of the human brain. *Neuron*, 72, 665–678.

Pray, L. (2008). Discovery of DNA structure and function: Watson and Crick. *Nature Education*, 1.

Putnam, H. (1975 [1967]). The nature of mental states. In: *Mind, Language and Reality*, Cambridge, MA: Cambridge University Press, pp. 429–440.

Raatikainen, P. (2010). Causation, exclusion, and the special sciences. *Erkenntnis*, 73, 349–363.

Ramachandran, V. S. & Hirstein, W. (1998). The perception of phantom limbs. *Brain*, 121, 1603–1630.

Reichenbach, H. (1956). *The Direction of Time*. Los Angeles, CA: University of Los Angeles Press.

Reutlinger, A. (2012). Getting rid of interventions. *Studies in History and Philosophy of Biological and Biomedical Sciences*, 43, 787–795.

Roediger, H. & McDermott, K. (1995). Creating false memories: Remembering words not presented in lists. *Journal of Experimental Psychology: Learning, Memory, and Cognition*, 21, 803–814.

Romero, F. (2015). Why there isn't interlevel causation in a mechanism. *Synthese*, 192, 3731–3755.

Rönnberg, J., Sönderfeldt, B., & Rinsberg, J. (2000). The cognitive neuroscience of signed language. *Acta Psychologica*, 105, 237–254.

Rugg, M. D. & Yonelinas, A. P. (2003). Human recognition memory: a cognitive neuroscience perspective. *Trends in Cognitive Sciences*, 7, 313–319.

Rumelhart, D. E., Smolensky, P., McClelland, J. L., & Hinton, G. E. (1986). Parallel distributed processing models of schemata and sequential thought processes. In: *Parallel Distributed Processing: Explorations in the Microstructure of Cognition*, Cambridge, MA: MIT Press, vol. 2, pp. 216–271.

Salmon, W. (1984). *Scientific Explanation and the Causal Structure of the World*. Princeton, NJ: Princeton University Press.

Salmon, W. (1989). *Four Decades of Scientific Explanation*. Minneapolis, MN: University of Minnesota Press.

Sandler, W. & Lillo-Martin, D. (2006). *Sign Language and Linguistic Universals*. Cambridge, NY: Cambridge University Press.

Saur, D., Kreher, B. W., Schnell, S., Kümmerer, D., Kellmeyer, P., Vry, M.-S., Umarova, R., Musso, M., Glauche, V., Abel, S., Huber, W., Rijntjes, M., Hennig, J., & Weiller, C. (2008). Ventral and dorsal pathways for language. *Proceedings of the National Academy of Sciences of the United States of America*, 105, 18035–18040.

Sauvage, M. (2010). ROC in animals: uncovering the neural substrates of recollection and familiarity in episodic recognition memory. *Consciousness and Cognition*, 19, 816–828.

Scholl, R. & Räz, T. (2013). Modeling causal structures: Volterra's struggle and Darwin's success. *European Journal of Philosophy of Science*, 3, 115–132.

Scoville, W. & Milner, B. (1957). Loss of recent memory after bilateral hippocampal lesions. *Journal of Neurology, Neurosurgery, and Psychiatry*, 20, 11–20.

Senior, C., Russell, T., & Gazzaniga, M. S. (eds.) (2006). *Methods in Mind*. Cambridge, MA: MIT Press.

Shapiro, L. (2010). Lessons from causal exclusion. *Philosophy and Phenomenological Research*, 81, 594–604.

Shapiro, L. & Polger, T. (2012). Identity, variability, and multiple realization in the special sciences. In: S. Gozzano & C. S. Hill (eds.), *New Perspectives on Type Identity: The Mental and the Physical*, Cambridge, MA: Cambridge University Press, pp. 264–287.

Shapiro, L. & Sober, E. (2007). Epiphenomenalism. The dos and don'ts. In: G. Wolters & P. Machamer (eds.), *Thinking About Causes: From Greek Philosophy to Modern Physics*, Pittsburgh, PA: University of Pittsburgh Press, pp. 235–264.

Silberstein, M. & Chemero, A. (2013). Constraints on localization and decomposition as explanatory strategies in the biological sciences. *Philosophy of Science*, 80, 958–970.

Silva, A., Bickle, J., & Landreth, A. (2013). *Engineering the Next Revolution in Neuroscience*. New York, NY: Oxford Univesity Press Neuroscience.

Skipper, R. A. & Millstein, R. A. (2005). Thinking about evolutionary mechanisms: Natural selection. *Studies in History and Philosophy of Biological and Biomedical Sciences*, 36, 327–347.

Smart, J. (1959). Sensations and brain processes. *Philosophical Review*, 68, 141–156.

Sodian, B. (2005). Theory of mind. The case for conceptual development. In: W. Schneider, R. Schumann-Hengsteler, & B. Sodian (eds.), *Young children's cognitive development. Interrelationships among working memory, theory of mind, and executive functions*, Hillsdale, NJ: Erlbaum, pp. 95–130.

Spirtes, P., Glymour, C., & Scheines, R. (1993). *Causation, Prediction and Search*. New York, NY: Springer.

Squire, L. R. (2004). Memory systems of the brain: A brief history and current perspective. *Neurobiology of Learning and Memory*, 82, 171–177.

Stark, C. E. L. & Squire, L. R. (2000). Functional magnetic resonance imaging (fMRI) activity in the hippocampal region during recognition memory. *Journal of Neuroscience*, 20, 7776–7781.

Stephan, A. (1999). Varieties of emergence. *Evolution and Cognition*, 5, 50–59.

Stokoe, W. C. (1960). *Sign language structure: an outline of the visual communication systems of the American deaf*. Studies in Linguistics: Occasional Papers. Buffalo: University of Buffalo.

Storbeck, J. & Clore, G. (2005). With sadness comes accuracy; with happiness, false memory. mood and the false memory effect. *Psychological Science*, 16, 785–791.

Strevens, M. (2008). Comments on Woodward, Making Things Happen. *Philosophy and Phenomenological Research*, 77, 171–192.

Sutton-Spence, R. & Woll, B. (1998). *The Linguistics of British Sign Language. An Introduction*. Cambridge, UK: Cambridge University Press.

Suzuki, M., Johnson, J. D., & Rugg, M. D. (2011). Recollection-related hippocampal activity during continuous recognition: A high-resolution fmri study. *Hippocampus*, 21, 575–583.

Tabery, J. (2004). Synthesizing activities and interactions in the concept of a mechanism. *Philosophy of Science*, 71, 1–15.

Teervort, B. (1973). Could there be a human sign language? *Semiotica*, 9, 347–382.

Thut, G. & Miniussi, C. (2009). New insights into rhythmic brain activity from TMS–EEG studies. *Trends in Cognitive Sciences*, 13, 182–188.

Tolman, E. (1948). Cognitive maps in rats and men. *Psychological Review*, 55, 189–208.

Tsien, J. Z., Huerta, P. T., & Tonegawa, S. (1996). The essential role of hippocampal CA1 NMDA receptor-dependent synaptic plasticity in spatial memory. *Cell*, 87, 1327–1338.

Turing, A. (1936). On computable numbers, with an application to the entscheidungsproblem. *Proceedings of the London Mathematical Society*, 2, 230–265.

Ungerleider, L. & Mishkin, M. (1982). Two cortical visual systems. In: D. Ingle, M. Goodale, & R. Mansfield (eds.), *Analysis of Visual Behavior*, Cambridge, MA: MIT Press, pp. 549–586.

Ungerlieder, L. & Haxby, J. (1994). 'What' and 'where' in the human brain. *Current Opinion in Neurobiology*, 4, 157–165.

van Riel, R. (2014). *The Concept of Reduction*. Dordrecht: Springer.

von Neumann, J. (1993 [1945]). First draft of a report on the EDVAC. *IEEE Annuals of the History of Computing*, 15, 27–75.

Ward, J. (2006). *The Student's Guide to Cognitive Neuroscience*. Sussex: Psychology Press.

Waters, K. (2007). Causes that make a difference. *Journal of Philosophy*, 54, 551–579.

Waters, K. (2014). Causes that matter in scientific practice. Presentation at Universität zu Köln, January 17th.

Watson, J. (1913). Psychology as the behaviorist views it. *Psychological Review*, 20, 158–177.

Watson, J. D. & Crick, F. H. C. (1953). A structure for deoxyribose nucleic acid. *Nature*, 171, 737–738.

Wernicke, C. (1874). *Der aphasische Symptomencomplex: eine psychologische Studie auf anatomischer Basis*. Breslau: Cohen & Weigart.

Wimsatt, W. (1972). Teleology and the logical structure of the function statements. *Studies in History and Philosophy of Science*, 3, 1–80.

Wimsatt, W. (1976). Reductionism, levels of organization, and the mind-body problem. In: G. G. Globus, G. Maxwell, & I. Savodnik (eds.), *Consciousness and the Brain: A Scientific and Philosophical Inquiry*, New York, NY: Plenum Press, pp. 205–267.

Wimsatt, W. (2007). *Re-Engineering Philosophy for Limited Beings. Piecewise Approximations to Reality*. Cambridge, MA: Harvard University Press.

Wittgenstein, L. (1974). Philosophical grammar. In: A. Kenny (trans.) & R. Rhees (eds.), *Philosophical Grammar*, Oxford: Blackwell.

Woodward, J. (2000). Explanation and invariance in the special sciences. *British Journal for the Philosophy of Science*, 51, 197–254.

Woodward, J. (2002). What is a mechanism? a counterfactual account. *Philosophy of Science*, 69, S366–S377.

Woodward, J. (2003). *Making Things Happen: A Theory of Causal Explanation*. New York: Oxford University Press.

Woodward, J. (2008a). Mental causation and neural mechanisms. In: J. Hohwy & J. Kallestrup (eds.), *Being Reduced: New Essays on Reduction, Explanation, and Causation*, Oxford: Oxford University Press, pp. 218–262.

Woodward, J. (2008b). Response to strevens. *Philosophy and Phenomenological Research*, Philosophy and Phenomenological Research 77, 193–212.

Woodward, J. (2013). Mechanistic explanation: Its scope and limits. *Proceedings of the Aristotelian Society*, Suppl. Vol. LXXXVII, 39–65.

Woodward, J. (2015). Interventionism and causal exclusion. *Philosophy and Phenomenological Research*, 91, 303–347.

Woodward, J. & Hitchcock, C. (2003). Explanatory generalizations, part 1: A counterfactual account. *Noûs*, 37, 1–24.

Woolsey, T. & Wann, J. (1976). Areal changes in mouse cortical barrels following vibrissa damage at different postnatal stages. *Journal of Comparative Neurology*, 170, 53–66.

Wundt, W. (1874). *Grundzüge der physiologischen Psychologie*. Leipzig: Engelmann.

Wundt, W. (1907). *Über Ausfrageexperimente und über die Methoden zur Psychologie des Denkens*, vol. 3 of *Psychologische Studien*. Leipzig: Engelmann.

Ylikoski, P. (2013). Causal and constitutive explanation compared. *Erkenntnis*, 78, 277–297.

Yonelinas, A. P. & Parks, C. M. (2007). Receiver operating characteristics (ROCs) in recognition memory: A review. *Psychological Bulletin*, 133, 800–832.

Zednik, C. & Jäkel, F. (2014). How does bayesian reverse-engineering work? Manuscript.

Key Terms

action potential: electrical impulse produced by a neuron

adaptation: getting used to, adjustment over time

aphasia: condition characterized by total or partial loss of the ability to communicate using spoken or written words

axon: outreaching projection of a nerve cell along which signals are propagated

bottom-up study: experiment interfering with lower-level processes to elicit higher-level effects

Brodmann areas: brain areas defined by cytoarchitectural organization and numbered by Korbinian Brodmann

C-M model: account according to which explanations describe causal processes

cognate: a sign common to BSL and SSL with the same meaning in both SLs

consolidation: process encoding memory

D-N model: account according to which explanation proceeds by nomological deduction

dissociation: a patient's selective functional impairment on a cognitive task, often associated with focal brain injury

DTI: diffusion tensor imaging; MRI technique used to visualize architectonical details

dyslexia: developmental condition that causes difficulty in learning to read or interpret words, letters, and other symbols

emergence: complex structures/properties arise from the combination of multiple simple ones

explanandum: observed phenomenon that needs to be explained

explanans: what does the explaining

familiarity: vague feeling of déjà-vu

fat-handed intervention: intervention that affects multiple variables at once

foreign sign: a sign not part of a signer's native SL

gel electrophoresis: method for separation and analysis of macromolecules such as DNA

gene knockout: genetic technique in which a target gene is made inoperative in an organism

iconicity: presence of a systematic visually motivated relation between a sign's referent and its phonological form

identity theory: theory according to which mental processes are identical to physical ones

interaction effect: combined effect of independent variables

interventionism: theory of causation and scientific (causal) explanation based on difference making

interventionist manifesto: "No causal difference without a difference in manipulability!"

klinotaxis: spatial orientation behavior in C. elegans

LTP: long-term memory; lasting strengthening of synaptic connections due to repeated synchronous stimulation

DOI 10.1515/9783110530940-017

main effect: overall effect of one variable on another
mechanism: organized set of entities and their activities (Xs' ϕ-ings)
mechanistic component: constitutively relevant part of a mechanism
mechanistic explanation: a phenomenon (S's ψ-ing) is explained by relevant organized entities and activities (Xs' ϕ-ings)
mechanistically mediated effect: hybrid of causal and constitutive relation
mental lexicon: a language user's knowledge of words and their meaning
mere interaction: experimental manipulation that does not qualify as Woodward-style intervention, i.e. is not a manipulation of some X with respect to some Y
multiple realizability: thesis that a mental state/process can be implemented by different physical states/processes

native sign: a sign forming part of a signer's native SL
non-sign: a phonologically implausible fake-sign carrying reduced phonological information

PCA: principal component analysis; data analysis method generating linear equations
phantom limb: sensation of an amputated limb as still there
phoneme: smallest perceptually distinct unit of speech
pseudo-interventions: way of interpreting mere interactions as if there had been interventions

recognition memory: ability to recognize previously encountered events, objects, or people
recollection: recognition based on episodic information
ruthless reductionism: account that accepts only complete molecular-level explanations

semantic memory: the locus of concepts, our knowledge of the world and its structure
single unit recording: method where intracranial electrodes record from neurons in cortex
SL phonology: sensorimotor characteristics of signs
spatial memory: memory required for spatial orientation in familiar environments
split brain: condition where communication between the cerebral hemispheres is impaired, associated with characteristic cognitive deficits
subtraction: data analysis method based on contrasts between groups or conditions
supervenience: ontological relation; if A supervenes on B there cannot be an A-difference without a B-difference
synapse: terminal nerve ending

ToM: theory of mind; ability to attribute mental states to others
TMS: transcranial magnetic stimulation; temporally "knocks out" a targeted brain area
top-down study: experiment interfering with higher-level processes to elicit lower-level effects
tracer: marked molecule, e.g. injected into cells to track physiological processes or connections
translational research: discoveries made in simple models/model organisms are transferred to complex target organisms (often in clinical settings)

voxel: volumetric pixel, cubic unit of space for which MRI measures data points

w-question: What would have happened if things had been different?

Index

www.ingramcontent.com/pod-product-compliance
Lightning Source LLC
Chambersburg PA
CBHW051957270326
41929CB00015B/2696